Mike Holt's Illustrated Guide to

POWER QUALITY

Mike Holt Enterprises, Inc.
888.NEC.CODE (632.2633) • www.MikeHolt.com • Info@MikeHolt.com

Revised Date: January 30, 2012

NOTICE TO THE READER

The publisher does not warrant or guarantee any of the products described herein or perform any independent analysis in connection with any of the product information contained herein. The publisher does not assume, and expressly disclaims, any obligation to obtain and include information other than that provided to it by the manufacturer.

The reader is expressly warned to consider and adopt all safety precautions that might be indicated by the activities herein and to avoid all potential hazards. By following the instructions contained herein, the reader willingly assumes all risks in connection with such instructions.

The publisher makes no representation or warranties of any kind, including but not limited to, the warranties of fitness for particular purpose or merchantability, nor are any such representations implied with respect to the material set forth herein, and the publisher takes no responsibility with respect to such material. The publisher shall not be liable for any special, consequential, or exemplary damages resulting, in whole or part, from the reader's use of, or reliance upon, this material.

Mike Holt's Illustrated Guide to Power Quality
Second Printing: January 2012

Technical Illustrator: Mike Culbreath
Cover Design: Madalina Iordache-Levay
Layout Design and Typesetting: Cathleen Kwas

COPYRIGHT © 2012 Charles Michael Holt
ISBN 978-1-932685-30-5

Produced and Printed in the USA

For more information, call 888.NEC.CODE (632.2633), or e-mail Info@MikeHolt.com.

All rights reserved. No part of this work covered by the copyright hereon may be reproduced or used in any form or by any means graphic, electronic, or mechanical, including photocopying, recording, taping, or information storage and retrieval systems without the written permission of the publisher. You can request permission to use material from this text by either calling 866.632.2633, e-mailing Info@MikeHolt.com, or visiting www.MikeHolt.com.

NEC®, NFPA 70®, NFPA 70E® and *National Electrical Code*® are registered trademarks of the National Fire Protection Association.

 This logo is a registered trademark of Mike Holt Enterprises, Inc.

If you are an instructor and would like to request an examination copy of this or other Mike Holt Publications:

Call: 888.NEC.CODE (632.2633) • Fax: 352.360.0983
E-mail: Info@MikeHolt.com • Visit: www.MikeHolt.com

You can download a sample PDF of all our publications by visiting www.MikeHolt.com

I dedicate this book to the
Lord Jesus Christ,
my mentor and teacher.
Proverbs 16:3

One Team

To Our Instructors and Students:

We're committed to providing you the finest product with the fewest errors, but we're realistic and know that there'll be errors found and reported after the printing of this book. The last thing we want is for you to have problems finding, communicating, or accessing this information. It's unacceptable to us for there to be even one error in our textbooks or answer keys. For this reason, we're asking you to work together with us as One Team.

Students: Please report any errors you may find to your instructor.

Instructors: Please communicate these errors to us by sending an e-mail to corrections@mikeholt.com.

Our Commitment:

We'll continue to list all of the corrections that come through for all of our textbooks and answer keys on our Website. The most up-to-date answer keys will always be available to instructors to download from our instructor Website. We don't want you to have problems finding this updated information, so we're outlining where to go for all of this below:

To view textbook and answer key corrections: Students and instructors go to our Website, www.MikeHolt.com, click on "Books" in the sidebar of links, and then click on "Corrections."

To download the most up-to-date answer keys: Instructors go to our Website, www.MikeHolt.com, click on "Instructors" in the sidebar of links and then click on "Answer Keys." On this page you'll find instructions for accessing and downloading these answer keys.

If you're not registered as an instructor you'll need to register. Your registration will be sent to our educational director who in turn will review and approve your registration. In your approval e-mail will be the login and password so you can have access to all of the answer keys. If you have a situation that needs immediate attention, please contact the office directly at 888.NEC.CODE (632.2633) or visit us online at www.MikeHolt.com.

Call 888.NEC.CODE (632.2633) or visit us online at www.MikeHolt.com

Table of Contents

About the Author vii
About the Graphic Illustrator viii

CHAPTER 1—INTRODUCTION TO POWER QUALITY 1
1.1 Scope of This Program 1
1.2 Changing Electrical Environment 1
1.3 Poor Power Quality 2
1.4 Why Power Quality Is Important 3
1.5 What's the Problem? 3

CHAPTER 2—ELECTRICAL THEORY 5
2.1 The Electrical Circuit 5
2.2 Electron Current Flow Theory 6
2.3 Ohm's Law 8
2.4 Circuit Voltage Distribution 8
Conclusion 8
Chapter 2 Review Questions 9

CHAPTER 3—ALTERNATING CURRENT 11
3.1 Alternating Current 11
3.2 Alternating-Current Generator 11
3.3 Sinusoidal Waveform 13
3.4 Sinusoidal Waveform Values 14
3.5 Peak 15
3.6 RMS 15
3.7 Zero Voltage Crossings 17
3.8 Nonsinusoidal Waveform 18
3.9 Crest Factor 18
3.10 Frequency 20
3.11 Phase 20
3.12 Out-of-Phase 21
3.13 Eddy Currents—Conductors 21
3.14 Skin Effect 22
3.15 Eddy Currents—Metal Parts 23
3.16 Hysteresis Loss 24
3.17 Inductive Heating 24
3.18 Capacitance 25
3.19 Inductive Reactance 26
3.20 Reactance 26
3.21 Resonance 27
Conclusion 29
Chapter 3 Review Questions 29

CHAPTER 4—NEUTRAL CONDUCTOR 33
4.1 Neutral Conductor 33
4.2 Neutral Current—2-Wire Circuit 33
4.3 Neutral 3-Wire, Single-Phase Circuit 34
4.4 Neutral Current—4-Wire Circuit 35
4.5 Skin Effect and Eddy Currents 37
4.6 Triplen Currents 37
4.7 Additive Triplen Currents on the Neutral Conductor 37
4.8 Neutral Conductor Size and Load Balance 38
Conclusion 38
Chapter 4 Review Questions 39

CHAPTER 5—HARMONICS 41
5.1 Linear Load 41
5.2 Nonlinear Load 42
5.3 Harmonics 43
5.4 Harmonic Order 43
5.5 Resultant Nonsinusoidal Waveform 45
5.6 Triplen Harmonics 45
5.7 Neutral Current—4-Wire Circuit 46
5.8 Positive Sequence Harmonic 46
5.9 Negative Sequence Harmonic 46
5.10 Total Harmonic Distortion (THD) 48
5.11 Harmonic Effects 48
Conclusion 49
Chapter 5 Review Questions 50

Table of Contents

CHAPTER 6—VOLTAGE DISTURBANCES ... 53
6.1 Voltage Distortion 53
6.2 Voltage Flat-Topping 54
6.3 Voltage Notching 54
6.4 Zero Voltage Crossings 55
6.5 Voltage Sags 56
6.6 Undervoltage 57
6.7 Transients (Voltage Spikes) 58
6.8 Surge Protection Devices 59
6.9 Voltage Swells 61
6.10 Overvoltage 61
6.11 Unbalanced Line Voltage 61
Conclusion 62
Chapter 6 Review Questions 63

CHAPTER 7—VOLTAGE WINDOW 65
7.1 Premises Voltage Window 65
7.2 Equipment Voltage Window 65
Conclusion 68
Chapter 7 Review Questions 68

CHAPTER 8—ELECTRICAL NOISE 69
8.1 Noise from Arcing at Terminals 69
8.2 Noise from Equipment 70
8.3 Metal Conduit Shielding 72
8.4 Shielding 72
Conclusion 73
Chapter 8 Review Questions 74

CHAPTER 9—GROUNDING AND BONDING 77
9.1 System Grounding 77
9.2 Ungrounded System 78
9.3 Solidly Grounded System 79

9.4 System Grounding 80
9.5 Resistance Grounded System 80
9.6 Equipment Grounding 82
9.7 Communications Grounding 85
9.8 Lightning Protection System 87
9.9 Objectionable Current 90
9.10 Electronic Equipment 92
Conclusion 92
Chapter 9 Review Questions 93

CHAPTER 10—POWER QUALITY ISSUES 95
10.1 Adjustable Speed Drives 95
10.2 Busway Failure 96
10.3 Capacitors 96
10.4 Conductor Failure 98
10.5 Circuit Breakers 99
10.6 Electric-Discharge Luminaires 100
10.7 Dimmers 100
10.8 Overcurrent Protection Devices 101
10.9 Generators 101
10.10 Laser Printers 103
10.11 Light Flicker 103
10.12 Modular Office Furniture 104
10.13 Motors 104
10.14 Panelboards 105
10.15 Raceways 106
10.16 Transformers 106
10.17 Relays 109
10.18 Transfer Switches 109
10.19 Uninterruptible Power Supplies (UPS) 109
Conclusion 110
Chapter 10 Review Questions 111

TECHNICAL REFERENCES 117

About the Author

Mike Holt

Mike Holt worked his way up through the electrical trade from an apprentice electrician to become one of the most recognized experts in the world as it relates to electrical power installations. He has worked as a journeyman electrician, master electrician, and electrical contractor. Mike's experience in the real world gives him a unique understanding of how the *NEC* relates to electrical installations from a practical standpoint. You'll find his writing style to be direct, non-technical, and practical.

Did you know that he didn't finish high school? So if you struggled in high school or if you didn't finish it at all, don't let this get you down, you're in good company. As a matter of fact, Mike Culbreath, Master Electrician, who produces the finest electrical graphics in the history of the electrical industry, didn't finish high school either. So two high school dropouts produced the text and graphics in this textbook! However, realizing success depends on one's continuing pursuit of education. Mike immediately attained his GED (as did Mike Culbreath) and ultimately attended the University of Miami's Graduate School for a Master's degree in Business Administration (MBA).

Mike Holt resides in Central Florida, is the father of seven children, and has many outside interests and activities. He's a six-time National Barefoot Water-Ski Champion (1988, 1999, 2005, 2006, 2007, and 2008), has set many national records, has competed in three World Championships (2006, 2008, and 2010) and continues to train and work out year-round so that he can qualify to ski in the 2012 World Barefoot Championships at the age of 61!

What sets him apart from some, is his commitment to living a balanced lifestyle; placing God first, family, career, then self.

Mike Holt—Special Acknowledgments

First, I want to thank God for my godly wife who's always by my side and my children, Belynda, Melissa, Autumn, Steven, Michael, Meghan, and Brittney.

A special thank you must be sent to the staff at the National Fire Protection Association (NFPA), publishers of the *NEC*—in particular Jeff Sargent for his assistance in answering my many *Code* questions over the years. Jeff, you're a "first class" guy, and I admire your dedication and commitment to helping others understand the *NEC*. Other former NFPA staff members I would like to thank include John Caloggero, Joe Ross, and Dick Murray for their help in the past.

A personal thank you goes to Sarina, my long-time friend and office manager. It's been wonderful working side-by-side with you for over 25 years nurturing this company's growth from its small beginnings.

About the Graphic Illustrator

Mike Culbreath

Mike Culbreath devoted his career to the electrical industry and worked his way up from an apprentice electrician to master electrician. While working as a journeyman electrician, he suffered a serious on-the-job knee injury. With a keen interest in continuing education for electricians, he completed courses at Mike Holt Enterprises, Inc. and then passed the exam to receive his Master Electrician's license. In 1986, after attending classes at Mike Holt Enterprises, Inc., he joined the staff to update material and later studied computer graphics and began illustrating Mike Holt's textbooks and magazine articles. He's worked with the company for over 25 years and, as Mike Holt has proudly acknowledged, has helped to transform his words and visions into lifelike graphics.

Mike Culbreath—Special Acknowledgments

I want to thank my wonderful children, Dawn and Mac, who have had to put up with me during the *Code* revision seasons.

I would like to thank Steve Arne, our amazing technical editorial director, Eric Stromberg, an electrical engineer and super geek (and I mean that in the most complimentary manner, this guy is brilliant), and Ryan Jackson, an outstanding and very knowledgeable *Code* guy, for helping me keep our graphics as technically correct as possible.

I also want to give a special thank you to Cathleen Kwas for making me look good with her outstanding layout design and typesetting skills. I would also like to acknowledge Belynda Holt Pinto, our Chief Operations Officer and the rest of the outstanding staff at Mike Holt Enterprises, for all the hard work they do to help produce and distribute these outstanding products.

And last but not least, I need to give a special thank you to Mike Holt for not firing me over 25 years ago when I "borrowed" one of his computers and took it home to begin the process of learning how to do computer illustrations. He gave me the opportunity and time needed to develop my computer graphic skills. He's been an amazing friend and mentor since I met him as a student many years ago. Thanks for believing in me and allowing me to be part of the Mike Holt Enterprises family.

Introduction to Power Quality

CHAPTER 1

1.1 Scope of This Program

Power quality is increasingly important to facility owners and/or managers, electrical equipment manufacturers, and end-users of electrical products.

Good power quality lowers electricity costs, minimizes maintenance problems with electrical installations in buildings and similar structures, and lets electrical utilization equipment operate properly and efficiently.

Poor power quality increases electricity costs, causes serious maintenance problems with electrical installations, leads to utilization equipment failures and operational problems, and can even create serious safety problems such as fires of electrical origin.

There are many factors that degrade power quality; but one of the most serious is harmonics, caused by what the *National Electrical Code* and other industry standards refer to as nonlinear loads.

Harmonics in electrical distribution systems, caused by nonlinear loads, affect everyone: people operating computers, electricians troubleshooting system problems, electrical contractors absorbing the cost of replacing damaged equipment, inspectors investigating the causes of electrical fires, and facilities management staff who are interested in effective and efficient equipment operation and reduced downtime.

The causes, effects, and prevention of harmonics are a complex subject, and few people in the electrical trade have a good working knowledge of the problem. This textbook provides the foundation necessary to understand the power quality and reliability issues and solutions, causes of harmonic problems, which solution(s) to apply, and when to call an expert.

1.2 Changing Electrical Environment

Much of today's electronic technology requires power that is sufficiently free of voltage and current disturbance so as not to cause undue energy waste or heating of supplied equipment. In decades past, this was not the case.

Notes

Notes

Several decades ago, most electrical equipment operated on ideal voltage and current waveforms and was not sensitive to power distortions. However, in the past 25 years (particularly since the late 1980s) there has been an explosion in the use of solid-state electronic technology. This highly efficient technology provides for improved product quality with increased productivity by the use of smaller and lighter electrical components. Today, we are able to manufacture products at costs substantially lower than in years past. However, this new technology requires clean electric power and is highly sensitive to power distortions such as high-voltage spikes and brownouts (undervoltage).

> **Author's Comment:** "Clean electric power" can be thought of as power that is sufficiently free of waveform distortion, voltage imbalance, induced noise, or other anomalies such that the combined anomalies do not cause undue energy waste or heating of supplied equipment. A "power distortion" is an undesirable change in the current or voltage waveform. Examples include phase shift and peak clipping. The most common example may be harmonic multiples riding on the fundamental waveform. The acceptable amount of power distortion varies with the application, but the mere presence of power distortion does not normally indicate a problem that needs to be solved.

New electronic devices contain rectifiers and capacitors that convert 60 Hz alternating current (ac) to direct current (dc) by the use of switching power supplies. In addition to converting alternating current to direct current, the current is sometimes converted back to alternating current, but at a different frequency. The conversion of alternating current to direct current by the use of rectifiers is the main cause of high harmonic currents in electrical distribution systems.

1.3 Poor Power Quality

Poor power quality increases operating costs, equipment malfunction and failure, and operational problems, among others. But most importantly, poor power quality can lead to serious safety issues.

Electronic equipment (switching power supplies) draws current differently than non-electronic equipment. Instead of a load having a constant impedance and drawing current in proportion to the sinusoidal voltage, electronic power supplies change their impedance by switching on and off near the peak of the voltage waveform. This switching on and off results in short, abrupt, nonsinusoidal current pulses during a controlled portion of the incoming peak voltage waveform.

These abrupt, pulsating currents introduce unanticipated reflective currents (harmonics) back into the power distribution system. These currents operate at frequencies other than the fundamental 60 Hz. Harmonic currents can be thought of as the vibration of water in a water line when a valve is rapidly opened and closed.

1.4 Why Power Quality Is Important

Distribution systems which are properly designed and installed should provide safe clean power, which reduces operating costs, minimizes maintenance issues, and improves employee productivity. Basically, this improves the bottom line for a business. How much downtime is acceptable? For reliability, it is important that there is dependable power available at all times. **Figure 1–1**

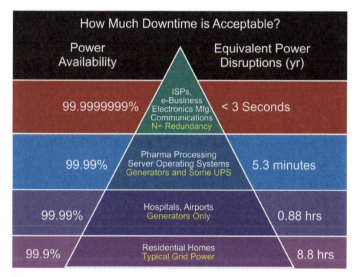

Figure 1–1

1.5 What's the Problem?

There are many factors that degrade power quality. The most significant include:

- Voltage, and current disturbances,
- Poor system design,
- Field wiring errors, and
- Improper grounding.

Notes

Electrical Theory

If it has been awhile since you studied electrical theory, the following chapters contain a short review of the most important concepts that need to be understood before you can understand harmonics. If you are comfortable with electrical theory, skip Chapter 2 and go directly to Chapter 3. You can always refer back to Chapter 2 if you do not understand an electrical concept in the remaining portions of this textbook. For a more complete understanding of electrical principles, see *Mike Holt's Illustrated Guide to Basic Electrical Theory* by Mike Holt.

Before we can begin to understand some of the technical issues relating to power quality, we need to do a short review of a few important electrical concepts and terms.

2.1 The Electrical Circuit

The movement of electrons for the production of work can be compared to the movement of water. Figure 2–1

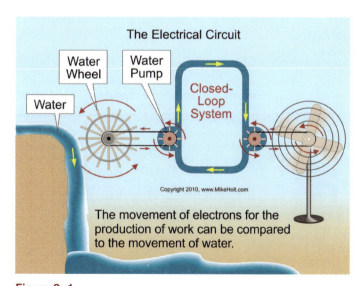

Figure 2–1

Notes

Notes

Electrical current flows from the power source through the load and then it returns to the power source. This path is called the "electrical circuit." **Figure 2–2**

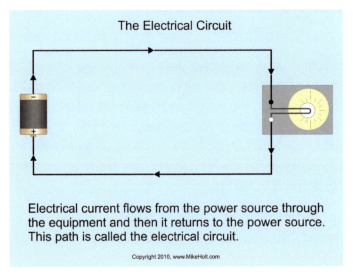

Figure 2–2

2.2 Electron Current Flow Theory

According to the electron current flow theory, electrons flow away from the negative terminal of the source, through the circuit and load, toward the positive terminal of the source. **Figure 2–3**

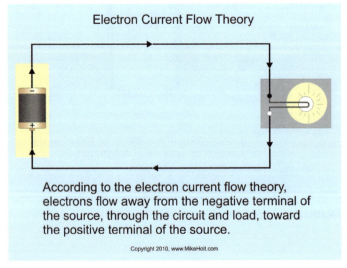

Figure 2–3

For current to flow, the power supply must apply sufficient pressure to cause the electrons to move through a closed loop from the source, through the load, and then back to the source. **Figure 2–4**

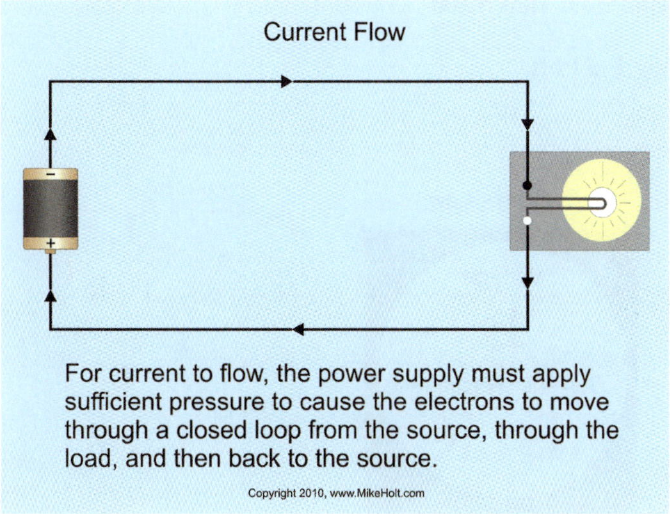

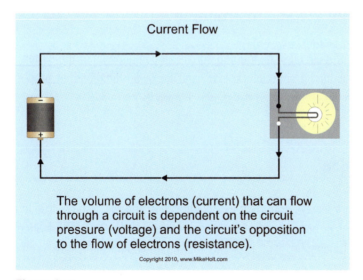

Figure 2–4

The volume of electrons (current) that can flow through a circuit is dependent on the circuit pressure (voltage) and the circuit's opposition to the flow of electrons (resistance). **Figure 2–5**

Figure 2–5

Notes

2.3 Ohm's Law

The relationship between voltage (E), current (I), and resistance (R) can be demonstrated via Ohm's Law. **Figure 2–6**

Ohm's Law: E = I x R

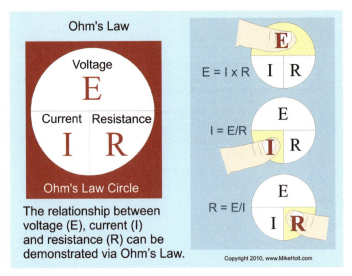

Figure 2–6

2.4 Circuit Voltage Distribution

The voltage of a circuit is distributed across all components of the circuit, such as the power supply, circuit conductors, load, and loose terminations, in proportion to their resistance to the entire circuit resistance. **Figure 2–7**

Conclusion

The basics are always important in any endeavor, and the same is true in understanding harmonics or any complicated electronics topic. This Chapter provided a review of some of the fundamentals of electricity to prepare you for the following chapters.

Electrical Theory — Chapter 2

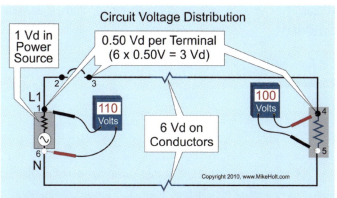

The voltage of a circuit is distributed across all components of the circuit, such as the power supply, circuit conductors, load, and loose terminations, in proportion to their resistance to the entire circuit resistance.

Figure 2–7

Chapter 2 Review Questions

1. Electrical current flows from the power source through the load and then it returns to the _____. This path is called the electrical circuit.

    (a) power source
    (b) earth
    (c) circuit
    (d) all of these

2. According to the _____ theory, electrons flow away from the negative terminal of the source, through the circuit and load, toward the positive terminal of the source.

    (a) power flow
    (b) electron current flow
    (c) big bang
    (d) source flow

Notes

3. The relationship between voltage (E), current (I), and resistance (R) can be demonstrated via _____ Law.

   (a) the Diminishing Return
   (b) Power Flow
   (c) Ohm's
   (d) Lenz's

4. The _____ of a circuit is distributed across all components of the circuit, such as the power supply, circuit conductors, load, and loose terminations, in proportion to their resistance to the entire circuit resistance.

   (a) voltage
   (b) parity
   (c) current
   (d) resonance

Alternating Current

3.1 Alternating Current

When a magnetic field moves through a conductor, the lines of force of the magnetic field cause electrons in the wire to flow in a specific direction. When the magnetic field moves in the opposite direction to the conductor, the electrons in the conductor flow in the opposite direction. Electrons only flow when there is relative motion between the conductor and the magnetic field. **Figure 3–1**

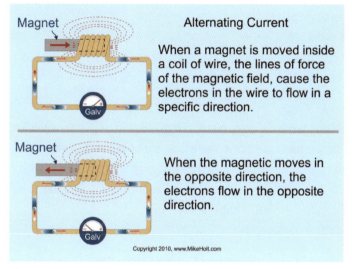

Figure 3–1

3.2 Alternating-Current Generator

A simple alternating-current generator consists of a loop of wire rotating between the lines of force between the opposite poles of a magnet. The halves of each conductor loop travel through the magnetic lines of force in opposite directions, causing the electrons within the conductor to move in a given direction. The magnitude of the voltage produced depends upon the number of turns of wire, the strength of the magnetic field, and the speed at which the coil rotates. **Figure 3–2**

Notes

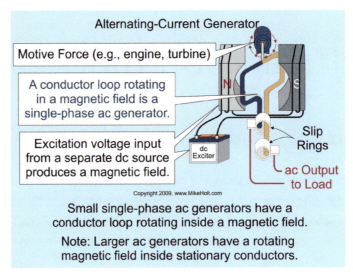

Figure 3–2

Author's Comment: The rotating conductor loop is called a "rotor" or "armature." Slip or collector rings and carbon brushes are used to connect the output voltage from the generator to an external circuit.

In generators that produce large quantities of electricity, the conductor coils are stationary and the magnetic field revolves within the coils. The magnetic field is produced by an electromagnet, instead of a permanent magnet. Using electromagnets permits the strength of the magnetic field (and thus the lines of force) to be modified, thereby controlling the output voltage. **Figure 3–3**

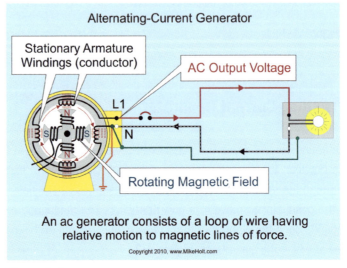

Figure 3–3

3.3 Sinusoidal Waveform

The waveform for alternating-current circuits is symmetrical, with positive above and negative below the zero reference level. For most alternating-current circuits, the waveform is called a "sine wave" or a "sinusoidal waveform." **Figure 3–4**

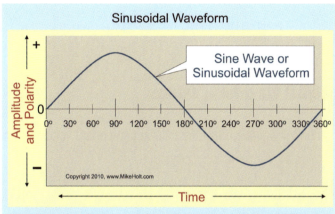

Figure 3–4

Figure 3–5 shows the relationship of the waveform and the rotor of a single-winding generator. The rotor of this simplified generator travels through a full 360 degrees during the generation of one sinusoidal waveform.

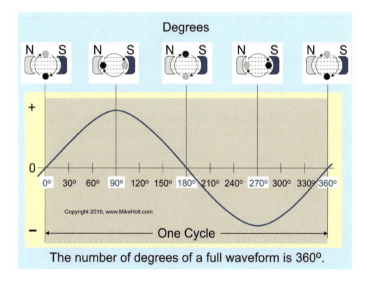

Figure 3–5

Notes

(1) The voltage starts at zero, when the rotor is not cutting any magnetic lines of force.

(2) As the rotor turns, the voltage increases from zero to a maximum value in one direction.

(3) It then decreases until it reaches zero.

(4) At zero, the voltage reverses polarity and increases until it reaches a maximum value at this opposite polarity.

(5) It decreases until it reaches zero again.

3.4 Sinusoidal Waveform Values

An alternating-current waveform has the following important values, **Figure 3–6**. They are:

(1) Peak voltage or current   Effective x 1.414

(2) Effective (True-RMS)   Peak x 0.707

(3) Instantaneous values

(4) Zero voltage crossings

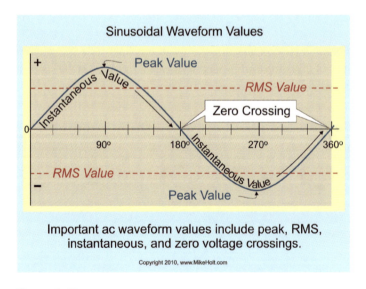

Figure 3–6

3.5 Peak

The Peak value of a sinusoidal waveform will be 1.40 times the Root Mean Square (RMS) value of the waveform. For example, the Peak voltage for a 120V RMS circuit is about 170V (120V x 1.40). **Figure 3–7**

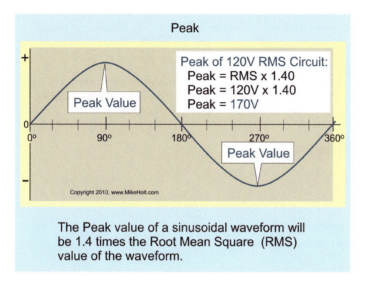

Figure 3–7

3.6 RMS

RMS is the square root of the mean of the squares of the instantaneous values. For a sinusoidal waveform, RMS values are less than Peak by a multiplier of 0.707, **Figure 3–8**. All electrical circuits are identified in RMS voltage, for example, 120/208V or 277/480V nominal. **Figure 3–9**

Notes

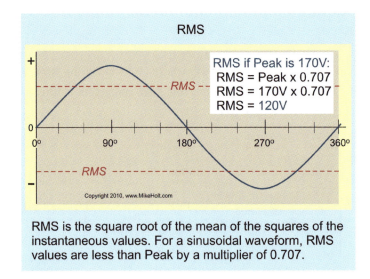

Figure 3–8

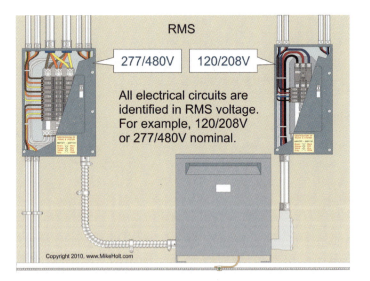

Figure 3–9

3.7 Zero Voltage Crossings

A sinusoidal voltage waveform has two zero voltage crossovers in each cycle. **Figure 3–10**

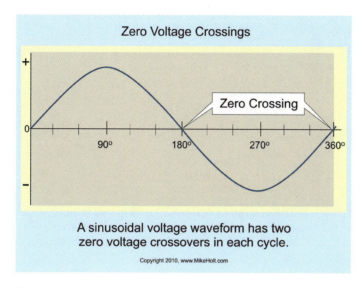

Figure 3–10

Large voltage notching, resulting from some large electronic loads, can create additional zero voltage crossings. **Figure 3–11**

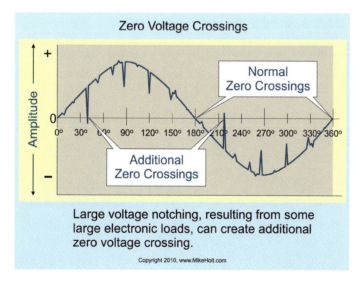

Figure 3–11

Notes

3.8 Nonsinusoidal Waveform

A "nonsinusoidal waveform" is any waveform that is not a sinusoidal shape, with positive above and negative below the zero reference level. Figure 3–12

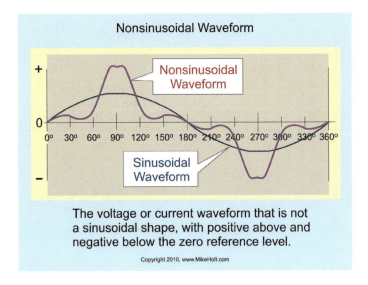

Figure 3–12

3.9 Crest Factor

The "crest factor" is the ratio between the peak value and the RMS value of a waveform. For example, the voltage crest factor of a sinusoidal load would be 1.40. The peak is 1.40 times greater than the RMS. Figure 3–13

Circuits supplying nonlinear loads typically have a current crest factor of 2.0 or greater. Figure 3–14

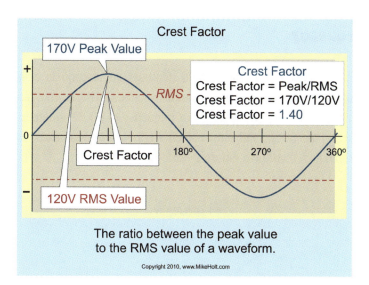

Figure 3–13

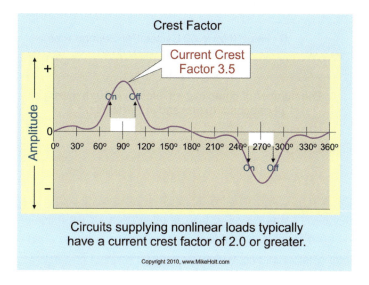

Figure 3–14

3.10 Frequency

The number of complete waveforms (360 degrees) per second is called the "frequency." Frequency is expressed as cycles per second, or Hertz (Hz). Most electrical power generated in the United States has a frequency of 60 Hz, whereas many other parts of the world use 50 Hz, and some use different power frequencies ranging from a low of 25 Hz to a high of 125 Hz. **Figure 3–15**

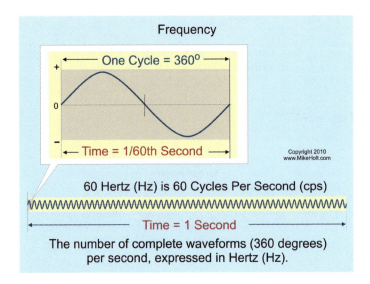

Figure 3–15

High-frequency electrical power of 415 Hz is often used for large computer systems and 400 Hz is used for airplane lighting. High-frequency power is often derived from motor-generator sets or other converters that use 60 Hz input power.

3.11 Phase

"Phase" is a term that indicates the time or degree relationship between two waveforms, such as voltage-to-current or voltage-to-voltage. When two waveforms are in step with each other, they are said to be "in-phase." In a purely resistive alternating-current circuit, the current and voltage are in-phase. This means that, at every instant, the current is exactly in step with the applied voltage. They both reach their zero and peak values at the same time, **Figure 3–16**

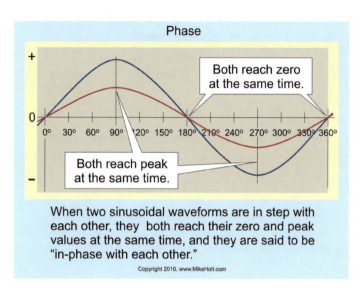

Figure 3–16

Phase differences are often expressed in degrees; one full waveform is equal to 360 degrees. For example, a three-phase generator has each of its windings out-of-phase with each other by 120 degrees.

3.12 Out-of-Phase

When waveforms are out of step with each other, they reach their zero and peak values at different times, and they are said to be "out-of-phase with each other." Three-phase power consists of three waveforms which are out-of-phase with each other by 120 degrees. Figure 3–17

3.13 Eddy Currents—Conductors

Electric currents induced in a conductor due to the expanding and collapsing alternating magnetic field within the conductor, are known as "eddy currents." Figure 3–18

Notes

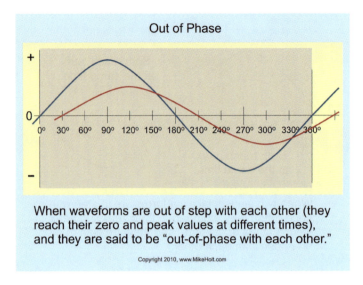

Figure 3–17

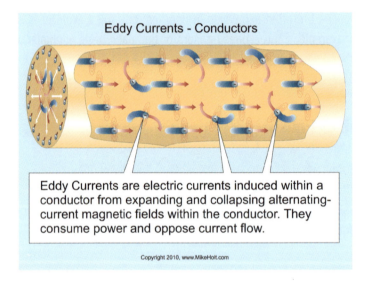

Figure 3–18

3.14 Skin Effect

Because eddy currents are greater in the center of a conductor, the applied current is forced to flow near the outer surface of the conductor. Skin effect increases with increasing alternating-current frequency. Figure 3–19

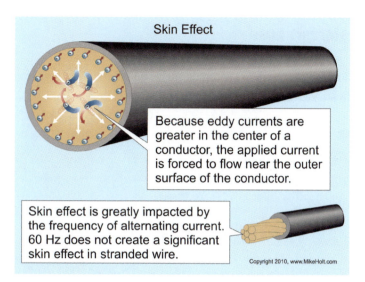

Figure 3–19

3.15 Eddy Currents—Metal Parts

Expanding and collapsing alternating magnetic fields passing through ferrous metals will introduce circulating eddy currents within the metal, Figure 3–20. Eddy currents in transformers can be reduced by using laminated cores.

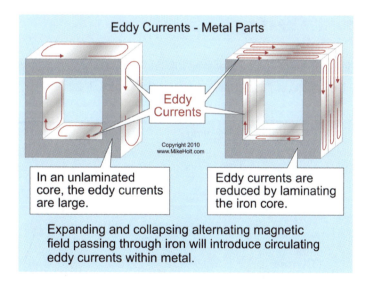

Figure 3–20

Notes

3.16 Hysteresis Loss

"Hysteresis loss" is the heating of ferrous metal parts when the iron molecules align and realign to an alternating magnetic field. **Figure 3–21**

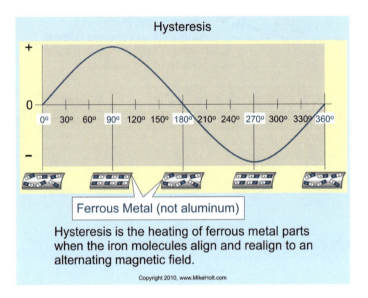

Figure 3–21

3.17 Inductive Heating

Heating of ferrous metal parts, because of eddy currents and hysteresis losses, increases by the square of the increase in frequency. **Figure 3–22**

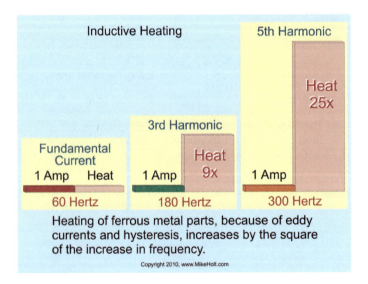

Figure 3–22

3.18 Capacitance

Capacitors are electrical devices that store energy in an electric field between closely spaced plates and are capable of releasing the energy at a later time. **Figure 3–23**

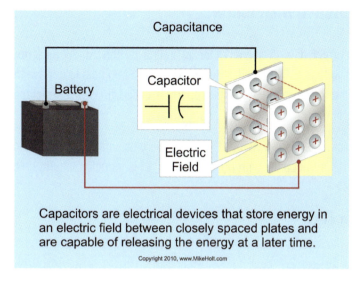

Figure 3–23

A capacitor is a device that resists changes in voltage. As the alternating-current sine wave reaches a positive peak, the capacitor fully charges to the same polarity. Then, as the current passes peak and decreases, the capacitor begins to discharge which has the effect of resisting the change of the alternating-current circuit voltage value. This results in what is called "capacitive reactance" and a shifting of the current waveform out-of-phase to the voltage waveform.

One way to think of this is that the capacitor responds to changes in current by increasing or decreasing its own amount of charge. Therefore, the voltage waveform change lags the current waveform change. The opposition offered to the flow of alternating current by a capacitor is called "capacitive reactance." This is expressed in ohms and abbreviated X_C.

Capacitive reactance is calculated using the following equation:

$X_C = 1/(2 \times \pi \times f \times C)$

Where Pi equals 3.14, "f" is the frequency in hertz, "C" is the capacitance in farads, and "X_C" is expressed in ohms.

Notes

A capacitor can be thought of as a device that resists changes in current. Because a capacitor introduces reactance to the circuit, it shifts the current waveform to lead the applied voltage by 90 degrees.

3.19 Inductive Reactance

Alternating-current flow in a conductor is limited by the conductor's resistance and self-induced voltage (CEMF). Self-induced voltage (CEMF) acts to oppose the change in current flowing in the conductor. This property is called "inductive reactance", and it is measured in ohms.

Inductive reactance is abbreviated X_L and can be calculated using the following equation, where "f" is frequency with units of hertz, "L" is inductance with units of Henrys, and π = 3.14.

$X_L = 2 \times π \times f \times L$

Author's Comment: Just remember that alternating current contains an additional element (reactance) that opposes the flow of electrons, besides conductor resistance.

In a purely inductive circuit, the CEMF waveform is 90 degrees out-of-phase with the circuit current waveform and 180 degrees out-of-phase with the applied voltage waveform. As a result, the applied voltage waveform leads the current waveform by 90 degrees.

Author's Comment: Just remember that an alternating-current circuit contains inductive reactance because the voltage and current are not in-phase with each other.

3.20 Reactance

"Reactance" is the property of resisting or impeding the flow of alternating current. It is measured in ohms and is denoted by the symbol "X."

Capacitive Reactance $X_C = 1/(2 \times π \times f \times C)$
Inductive Reactance $X_L = 2 \times π \times f \times L$

As frequency increases, X_L increases and X_C decreases. At a certain frequency they can reach an equal value, and at this point they are said to be in resonance.

3.21 Resonance

Resonance occurs when inductive reactance and capacitive reactance are equal ($X_L = X_C$).

Series LC Circuits

Series inductive capacitance circuits (LC) permit current to easily flow at a given frequency, and oppose or reduce current flow at all other frequencies. **Figure 3–24**

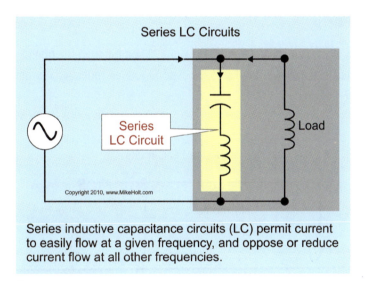

Figure 3–24

In a series LC circuit, as the frequency approaches the resonance frequency, the impedance of the LC circuit decreases so that the only opposition to current flow will be the conductor resistance. **Figure 3–25**

Notes

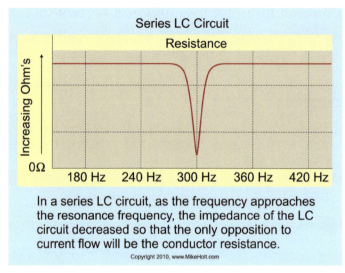

Figure 3–25

Series LC Filters

Passive inductive-capacitive series LC harmonic filters tuned to the offending harmonic and connected in parallel to the load, are used to reduce reflective harmonic currents.
Figure 3–26

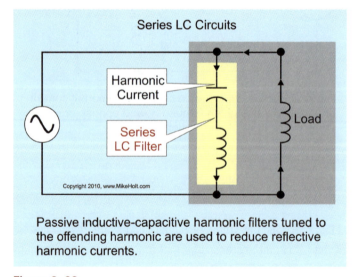

Figure 3–26

Conclusion

Alternating current has special characteristics that differ from direct current. This Chapter covered some of the important terms and concepts that are characteristics of alternating current, and are important concepts in regard to power quality issues.

Chapter 3 Review Questions

1. Electrons will flow through a conductor in a generator when there is relative motion between the conductor and a magnetic field.

   (a) True
   (b) False

2. A simple _____ consists of a loop of wire rotating between the lines of force between the opposite poles of a magnet.

   (a) conductor
   (b) ac generator
   (c) magnetic field
   (d) magnet

3. A generator's magnetic field can be produced by a permanent magnet.

   (a) True
   (b) False

4. For most alternating-current circuits, the waveform is called a _____.

   (a) sine wave
   (b) sinusoidal waveform
   (c) soidal wave
   (d) both a and b

5. Frequency is expressed as cycles per second, or Hertz (Hz).

   (a) True
   (b) False

Notes

Chapter 3 — Alternating Current

Notes

6. One full waveform is equal to _____ degrees.

    (a) 90
    (b) 120
    (c) 180
    (d) 360

7. Three-phase power consists of three waveforms which are out-of-phase with each other by _____ degrees.

    (a) 90
    (b) 120
    (c) 180
    (d) 360

8. Devices that intentionally introduce capacitance into circuits are called _____.

    (a) capacitors
    (b) condensers
    (c) components
    (d) both a and b

9. A capacitor can be thought of as a device that _____ changes in current.

    (a) resists
    (b) promotes
    (c) has no effect on.
    (d) all of these

10. The voltage induced within the conductor caused by its own expanding and collapsing magnetic field is known as _____.

    (a) applied current
    (b) self-induced voltage
    (c) magnetic flux
    (d) induced current

11. The induced voltage in a conductor carrying alternating current always opposes the change in current flowing through the conductor.

    (a) True
    (b) False

12. As frequency _____, X_L increases and X_C decreases. At a certain frequency they can reach an equal value, and at this point they are said to be in resonance.

    (a) increases
    (b) decreases
    (c) reaches zero Hz
    (d) both b and c

13. _____ is a condition that occurs when inductive reactance and capacitive reactance are equal ($X_L = X_C$).

    (a) Resonance
    (b) Impedance
    (c) Reactance
    (d) Induction

14. In a series LC circuit at resonance, since reactance is zero, the only opposition to current flow is the conductor resistance.

    (a) True
    (b) False

15. Inductive reactance (X_L) of a circuit increases with increased frequency, and capacitive reactance increases with increased frequency.

    (a) True
    (b) False

Notes

CHAPTER 4

Neutral Conductor

4.1 Neutral Conductor

The electrical term "neutral point" refers to the common point on a wye connection in a polyphase system or the midpoint on a single-phase portion of a three-phase delta system, or the midpoint of a 3-wire, single-phase system. The *National Electrical Code* calls the conductor connected to the neutral point of the power supply, which is intended to carry neutral current, the "neutral conductor." Figure 4–1

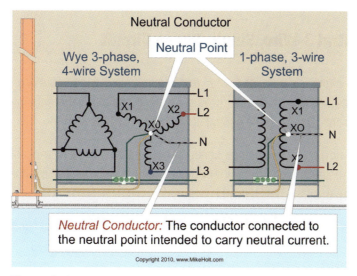

Figure 4–1

4.2 Neutral Current—2-Wire Circuit

In a simple 2-wire circuit, the magnitude of the current in the neutral will be the same as the current on the ungrounded conductor. Figure 4–2

Notes

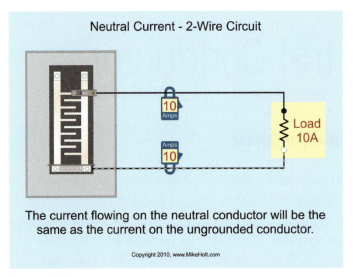

Figure 4–2

4.3 Neutral 3-Wire, Single-Phase Circuit

A multiwire circuit consists of two or more ungrounded conductors having a potential difference between them, and having an equal potential difference between each ungrounded conductor and the neutral conductor.

The current flowing on the neutral conductor of a three-wire, single-phase circuit will be the difference between the current flowing on the ungrounded conductors, and is often referred to as the unbalanced current.

The currents cancel because the currents flowing through the neutral conductor at any instant from the ungrounded conductors oppose each other. **Figure 4–3**

> CAUTION: If the ungrounded conductors of a multiwire circuit are connected to the same phase, the current on the neutral conductor will not cancel, but will add, which can cause an overload on the neutral conductor. **Figure 4–4**

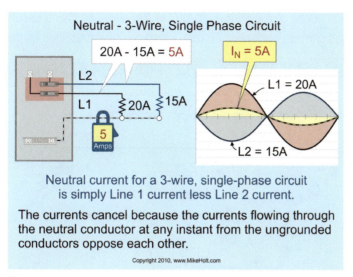

Figure 4–3

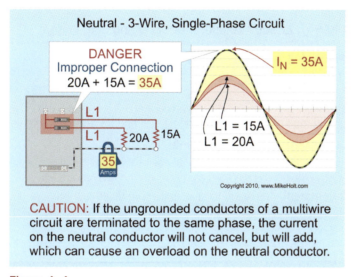

Figure 4–4

4.4 Neutral Current—4-Wire Circuit

The neutral conductor of a 4-wire, 120/208V or 277/480V circuit that supplies linear loads only carries the unbalanced phase current. If the current on all three phases is the same, then the neutral current is zero. In all other cases there will be neutral current.

Chapter 4 — Neutral Conductor

Notes

The neutral conductor of a 3-wire, 120/208V or 277/480V multiwire circuit of a 4-wire system always carries unbalanced current, which can be determined by the formula:

$$I_{Neutral} = \sqrt{(L1^2 + L2^2 + L3^2) - [(L1 \times L2) + (L2 \times L3) + (L1 \times L3)]}$$

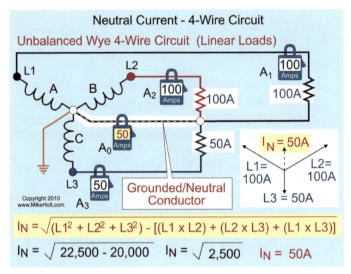

Figure 4–5

Where $I_{Neutral}$ is the total neutral current, and L1, L2, and L3 represent the line-to-neutral current of each phase conductor. **Figure 4–5**

Author's Comment: The same formula is used to determine the wye neutral current of a 3-wire multiwire circuit and 4-wire multiwire circuit.

Multiwire 120V and 277V branch circuits are more efficient and more cost-effective than multiple 2-wire 120V and 277V branch circuits. This is because a multiwire branch circuit permits a reduction in the total number of conductors required for the multiple line-to-neutral load, a smaller raceway size (fewer conductors), and at times, a reduction in circuit voltage drop. However, improper wiring or mishandling of the phase conductors can cause the neutral conductor to be overloaded. In addition, if the neutral conductor is opened, equipment can be destroyed because of overvoltages and undervoltages resulting from an open neutral condition.

4.5 Skin Effect and Eddy Currents

Eddy Currents—Neutral conductors and their associated terminals can overheat and fail because of eddy currents and skin effect. Eddy currents, although minimal cause additional heating because of the small independent currents produced as a result of the conductor's expanding and collapsing magnetic field.

Skin Effect—Skin effect is the phenomenon of increasing conductor resistance to higher-frequency currents, which has the effect of decreasing the effective cross sectional area of the conductor for current flow.

In addition to excessive neutral current and skin effect, conductors have a natural opposition to current flow (resistance), which causes the conductor to generate heat.

One final note—As a conductor or terminal heats up, the resistance increases one percent for each 2.50 percent increase in temperature. The combination of all these factors contributes to the failure of neutral conductors and their associated terminals.

4.6 Triplen Currents

Neutral 60 Hz currents from 4-wire balanced circuits that supply linear loads will cancel in the neutral conductor. However, when nonlinear loads are supplied from 4-wire wye circuits, odd multiples of 3rd order harmonic currents do not cancel, but instead, add together in the neutral conductor. Odd multiples of the 3rd order harmonics are called triplen harmonic currents; they are the 3rd, 9th, 15th, 21st, and so forth.

4.7 Additive Triplen Currents on the Neutral Conductor

Triplen harmonic currents on three-phase, 4-wire, Wye systems are in-phase with each other, and thus add arithmetically on the neutral conductor. Instead of A, B, and C-phase neutral currents canceling, the triplen harmonic currents add together with the 60-cycle noncancelled currents. The theoretical maximum value of neutral current can be three times the phase current, but rarely exceeds two times.

Notes

Notes

4.8 Neutral Conductor Size and Load Balance

Excessive neutral currents cause excessive temperatures, resulting in insulation failure on both the neutral conductor and adjacent phase conductors. Because the neutral conductor can be overloaded by unbalanced neutral currents and additive triplen harmonic currents, it is critical that the neutral conductor be sized properly.

This can be achieved by balancing loads of different phases as closely as possible, as well as by sizing the neutral conductor larger than the phase conductors where nonlinear loads are known to exist. Running a separate neutral conductor for each circuit instead of using multiwire circuits is also an option.

Danger: Failures of neutral conductors or terminals can cause the grounded neutral of multiwire circuits to open, resulting in fires and equipment failure because of overvoltage.

Conclusion

In general, overloading of the neutral conductor is a result of the improper wiring of the ungrounded conductors. Phase conductors connected to the same line potential while sharing a neutral conductor is one example of improper wiring. Such a connection causes the neutral conductor to carry the combined current of the phase conductors. Improper wiring of the neutral conductor can cause unstable equipment voltages, which can result in equipment failure and fires. The neutral conductor maintains and stabilizes the voltage of each circuit. If the neutral conductor common to more than one phase conductor is opened, the circuits change from parallel to series-parallel, which causes erratic voltage distribution.

Chapter 4 Review Questions

1. The electrical term _____ refers to the common point on a wye-connection in a polyphase system or the midpoint on a single-phase portion of a three-phase delta system, or the midpoint of a 3-wire, single-phase system.

    (a) neutral point
    (b) neutral conductor
    (c) unbalanced current
    (d) balanced current

2. A _____ consists of two or more ungrounded conductors having a potential difference between them, and having an equal potential difference between each ungrounded conductor and the neutral conductor.

    (a) linear load
    (b) multiwire circuit
    (c) unbalanced phase current
    (d) line-to-neutral current

3. When current flows on the neutral conductor of a multiwire circuit, the current is called the _____.

    (a) neutral point
    (b) neutral conductor
    (c) unbalanced current
    (d) balanced current

4. The neutral current on a balanced 4-wire circuit for linear loads equals _____.

    (a) 0A
    (b) 24A
    (c) 330A
    (d) 500A

5. In a balanced 4-wire multiwire circuit on a wye-connected system that supplies linear loads, the neutral conductor does not carry any current.

    (a) True
    (b) False

Chapter 4 — Neutral Conductor

Notes

6. The problem of neutral conductor overload due to harmonic currents can be solved by _____.

   (a) installing a larger neutral conductor
   (b) reducing the loads
   (c) running a separate neutral conductor for each phase
   (d) all of these

CHAPTER 5

Harmonics

Chapter 3 covered basic principles of alternating-current circuits and systems. This Chapter builds on that knowledge by introducing the concepts of harmonics caused by nonlinear loads. It defines nonlinear loads, explains what types of electrical equipment constitute nonlinear loads, and discusses the types of harmonics and other power quality problems they introduce into building power systems.

To understand harmonics, it's necessary to understand the basic difference between linear and nonlinear loads, as well as some other definitions.

5.1 Linear Load

A "linear load" is one that opposes the applied voltage with constant impedance, resulting in a current waveform that changes in direct proportion to the sinusoidal change in the applied voltage. Examples of these loads are resistance heating, incandescent lighting, and motors (to some extent). If the impedance is constant, and the applied voltage is sinusoidal, then the current and its waveform will also be sinusoidal. Figure 5–1

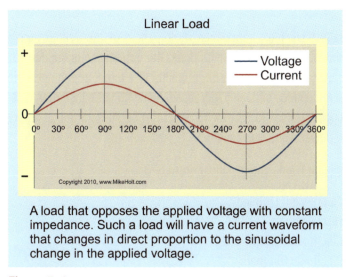

A load that opposes the applied voltage with constant impedance. Such a load will have a current waveform that changes in direct proportion to the sinusoidal change in the applied voltage.

Figure 5–1

Notes

Notes

5.2 Nonlinear Load

A "nonlinear load" is one that does not oppose the applied voltage with constant impedance. This results in a nonsinusoidal-current waveform. The nonsinusoidal-current waveform does not conform to the waveform of the applied voltage. Nonlinear loads change their load impedance on and off near the peak of the voltage waveform resulting in short, abrupt, high-current pulses. **Figure 5–2**

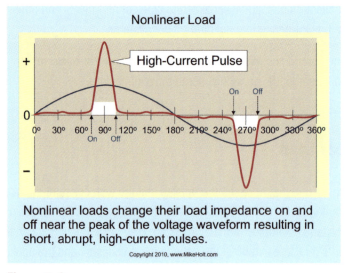

Figure 5–2

Nonlinear loads have high impedance during most of the voltage waveform. When the voltage waveform is at or near its peak, the impedance is suddenly reduced. The reduced impedance at the peak voltage results in a large, and sudden, rise in current flow until the impedance suddenly increases, resulting in an instantaneous drop in current. These nonsinusoidal, rapidly switched, current pulses introduce reflective currents back into the power distribution system.

Types of Nonlinear Loads can include: **Figure 5–3**

- Fluorescent Lighting
- HID Lighting
- Electronic Ballasts
- Electronic Dimmers
- Motors with Variable Frequency Drives
- Computers
- Uninterrupted Power Supplies (UPS)
- Laser Printers

- Data-Processing Equipment
- Other Electronic Equipment

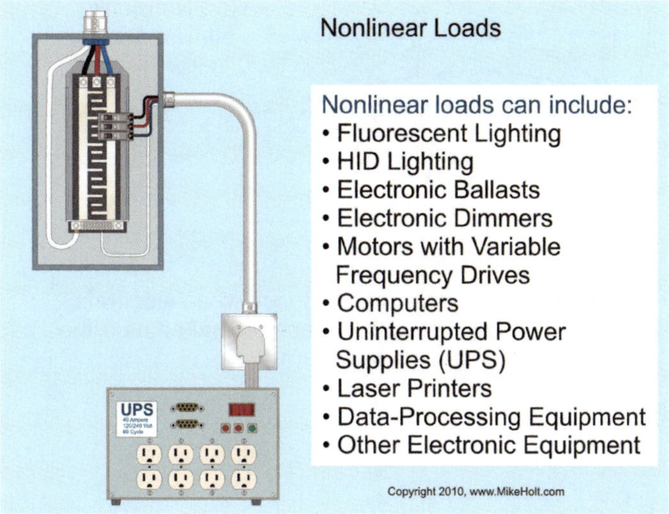

Figure 5–3

The *NEC* also provides a noncomprehensive list of items in the Note to the Article 100 definition of "Nonlinear Load."

5.3 Harmonics

Harmonic currents are produced when electronic loads change their impedance so the waveform of the current does not look like the voltage waveform. **Figure 5–4**

5.4 Harmonic Order

"Harmonic" is a term that describes waveforms that cycle at a frequency that is a multiple of the fundamental frequency. In common electrical distribution systems in the United States, the fundamental frequency is 60 Hz. When a current, or voltage, operates at other than the fundamental frequency of 60 Hz, it is said to operate at a specific harmonic order. The harmonic order is simply the ratio of the frequency of the harmonic to the fundamental frequency. **Figure 5–5**

Chapter 5 | **Harmonics**

Notes

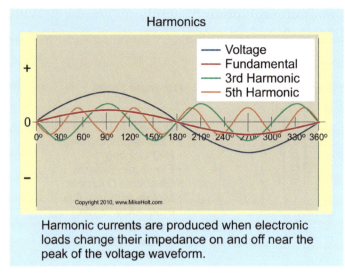

Harmonic currents are produced when electronic loads change their impedance on and off near the peak of the voltage waveform.

Figure 5–4

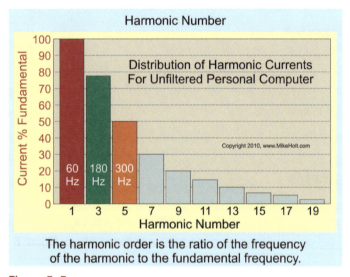

The harmonic order is the ratio of the frequency of the harmonic to the fundamental frequency.

Figure 5–5

Harmonic Order	
Harmonic Order	Frequency
Fundamental	60 Hertz
3rd Harmonic	180 Hertz
5th Harmonic	300 Hertz
11th Harmonic	660 Hertz

5.5 Resultant Nonsinusoidal Waveform

The combination of the fundamental 60 Hz frequency and reflective harmonic currents produce a nonsinusoidal resultant or complex current waveform. **Figure 5–6**

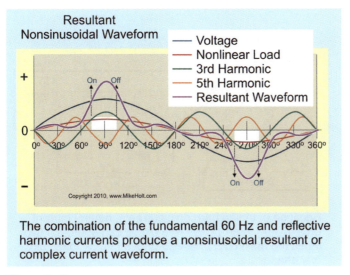

Figure 5–6

5.6 Triplen Harmonics

Single-phase nonlinear loads produce odd harmonics. Current which are multiples of the 3rd harmonic, are called "triplen" harmonics, for instance the 3rd, 9th, 15th, and so on. **Figure 5–7**

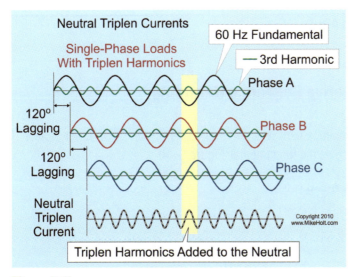

Figure 5–7

Notes

5.7 Neutral Current—4-Wire Circuit

Because of triplen harmonic currents, neutral current on a 4-wire, three-phase circuit can approach two times the maximum phase-to-neutral loads of any phase. **Figure 5–8**

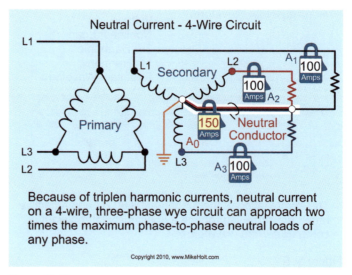

Figure 5–8

5.8 Positive Sequence Harmonic

Positive sequence harmonic currents of the 7th, 13th, and 19th order "rotate" in the same sequence as the fundamental: **Figure 5–9**

A-B-C to A-B-C

5.9 Negative Sequence Harmonic

Negative sequence harmonic currents of the 5th and 11th order "rotate" in the opposite sequence as the fundamental: **Figure 5–10**

C-B-A to A-B-C

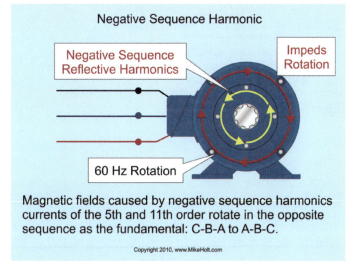

Figure 5–9

Figure 5–10

Notes

5.10 Total Harmonic Distortion (THD)

"Total harmonic distortion" is defined as the ratio of the harmonic voltage or current components to the fundamental; expressed as a percent.

THD = Harmonic/Fundamental x 100% Figure 5–11

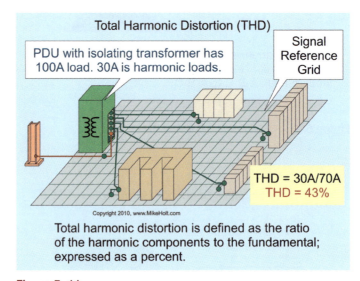

Figure 5–11

Waveform Content. Combining the fundamental (60 Hz) and reflective harmonic currents in the same conductor result in complex, nonsinusoidal-current waveforms. The complex waveform consists of the amplitude of the fundamental current and the sinusoidal amplitudes of the various harmonic currents at any instant in time. Complex waveforms may not resemble the fundamental waveform, or any of the individual harmonic waveforms.

5.11 Harmonic Effects

Because reflective harmonic currents operate at a higher frequency than the fundamental frequency, they have significant effects within the electrical distribution system. These include:

- Inductive heating of transformers, generators, and other electromagnetic devices, such as motors, relays and coils (because of increased eddy currents), skin effect, and hysteresis losses.

- Heating of conductors, breakers, fuses, and all other devices that carry current, because of increased I^2R heating due to eddy currents and skin effect.

- Inductive heating of ferrous metal (iron or steel) raceways, metal enclosures, and other parts, because of eddy currents and hysteresis losses.

- Unpredictable equipment operation, because of voltage distortion due to harmonic current.

- Equipment overheating or failure, because of additive harmonic currents, excessive voltage drop, and voltage distortion.

The actual harmonic content of any building varies depending on the types and number of installed loads that produce harmonics. Most building power systems can withstand nonlinear loads of up to 15 percent of the total electrical system capacity without concern, but when the nonlinear loads exceed 15 percent, some negative consequences begin to appear. For buildings that have nonlinear loading of more than 25 percent, particular problems that require correction become apparent.

Conclusion

This Chapter introduced the concept of harmonics caused by nonlinear loads. It defined nonlinear loads, explained what types of electrical equipment constitute nonlinear loads, and discussed the types of high-frequency signals and other problems they introduce into building power systems. It introduced a number of unfamiliar terms including "harmonic order," "triplen harmonics," and "total harmonic distortion (THD)."

Many types of electrical equipment can be classed as nonlinear loads. While some, such as arc welders, have been used for decades, the rapid increase in newer types of electronic equipment including personal computers, laser printers, electronic ballasts and dimmers, and adjustable speed drives have dramatically increased both the generation of harmonics and performance problems attributable to them.

Harmonic currents from nonlinear loads can be reduced at the source (switching power supply) with harmonic LC filters. But in general, nonlinear equipment does not have harmonic filters because there is no financial incentive for equipment manufacturers to provide this feature. Since harmonic currents are here to stay, we must adjust our thinking on electrical system design, installation, inspection, and maintenance. Designers and engineers must anticipate the inapparent overload of the electrical system, and the associated distortions to the voltage waveforms.

Notes

Chapter 5 Review Questions

1. A nonlinear load is one that opposes the applied voltage with constant impedance, resulting in a current waveform that changes in direct proportion to the sinusoidal change in the applied voltage.

    (a) True
    (b) False

2. _____ is an example of a linear load.

    (a) Resistance heating
    (b) Incandescent lighting
    (c) Electronic ballasts
    (d) both a and b

3. A nonlinear load is one that does not oppose the applied voltage with constant impedance. This results in a _____ current waveform

    (a) nonsinusoidal
    (b) sinusoidal
    (c) geometric
    (d) logarithmic

4. A linear load is one that opposes the applied voltage with constant impedance.

    (a) True
    (b) False

5. Harmonic currents are produced when electronic loads change their impedance so the waveform of the current does not look like the _____ waveform.

    (a) dc
    (b) voltage
    (c) resonance
    (d) intermittent

Harmonics | Chapter 5

6. The harmonic order is simply the ratio of the _____ of the harmonic to the fundamental frequency.

   (a) reflection
   (b) power
   (c) frequency
   (d) linear

7. In common electrical distribution systems in the United States, the fundamental frequency is _____.

   (a) 60 Hz
   (b) 30 Hz
   (c) 20 Hz
   (d) 10 Hz

8. The combination of the fundamental 60 Hz and reflective harmonic currents produce a nonsinusoidal _____ or complex current waveform.

   (a) divisor
   (b) resultant
   (c) fundamental
   (d) audio

9. Single-phase nonlinear loads produce odd harmonics. Current which are multiples of the 3rd harmonic, are called triplen harmonics. For instance, _____.

   (a) 3rd
   (b) 9th
   (c) 15th
   (d) all of these

10. _____ is defined as the ratio of the harmonic voltage or current components to the fundamental; expressed as a percent.

    (a) Parmonics
    (b) Reactance
    (c) Resonance
    (d) Total harmonic distortion

Notes

Notes

CHAPTER 6

Voltage Disturbances

Electronic devices change their impedance by switching on and off near the peak of the voltage waveform. Short, abrupt, high-current pulses during a controlled portion of the incoming peak voltage waveform are the result. The sudden rise of the pulsating current, reacting to the changing impedance of the nonlinear equipment, results in distorted nonsinusoidal-current waveforms. This nonsinusoidal current, when caused to flow through the power distribution transformer, produces a distorted voltage waveform.

When a distorted nonsinusoidal-voltage waveform is applied to a linear load, the resulting current has a distorted waveform. Resistive linear loads supplied by distorted voltage waveforms will operate without any particular problems. However, the effects of distorted nonsinusoidal-voltage waveforms can affect other types of equipment, such as UPSs, transformers, motors, generators, power factor correction capacitors, and power conditioning equipment. Sensitive electronic equipment (nonlinear loads) can fail to operate properly because of the voltage distortion characteristics of the distribution system, such as flat-topping, transients (notching and spikes), or erratic alternating-current voltage zero crossings.

6.1 Voltage Distortion

Sinusoidal voltage is produced at the source, but it is often degraded because of load variations that distort the voltage waveform by:

- Flat-Topping
- Notching
- Extra Zero Crossings
- Spikes/Transients
- Surges
- Sags
- Overvoltage
- Undervoltage

Notes

Notes

6.2 Voltage Flat-Topping

Voltage flat-topping can occur when nonlinear loads switch on and off near the peak of the voltage waveform. The voltage peak is thus depressed which can cause problems with some types of loads, for which voltage flat-topping results in a direct-current type of voltage supply lacking some of the alternating-current voltage characteristics. **Figure 6–1**

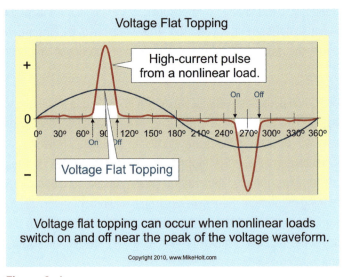

Figure 6–1

6.3 Voltage Notching

Silicon Controlled Rectifiers (SCR) used for three-phase motor variable speed drives, large UPS systems, or induction heating equipment can cause "notches" in the voltage waveform, **Figure 6–2**. Voltage notching results in additional zero crossings of the alternating-current sine wave.

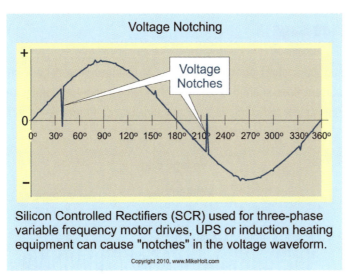

Figure 6–2

6.4 Zero Voltage Crossings

Large voltage notching, resulting from some larger electronic switched-mode power supplies, can create additional zero voltage crossings. This can create noise that may disrupt electronic control circuits. On large power systems, problems can be experienced when trying to synchronize standby power systems because it is more difficult to recognize the waveform. Figure 6–3

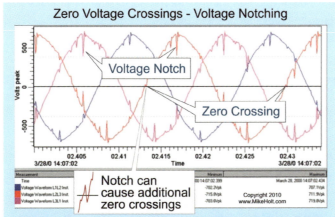

Figure 6–3

Notes

6.5 Voltage Sags

A "voltage sag" is a short duration decrease in voltage caused by a short duration increase in current flow, such as a motor starting. Sags can be of such short duration that the source is extremely difficult to locate and correct. Figure 6–4

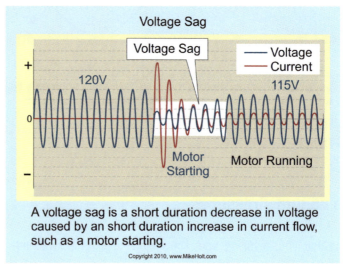

Figure 6–4

A voltage sag can also occur because of utility short circuits or ground faults. This is a very common problem and can hamper the proper operation of electronic equipment, including variable frequency drives. Figure 6–5

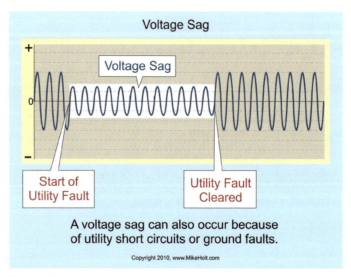

Figure 6–5

6.6 Undervoltage

A voltage sag lasting more than three seconds is classified as an undervoltage condition. The most common causes of undervoltage are excessive conductor voltage drop, an open neutral on a multiwire circuit, and loose terminals. **Figure 6–6**

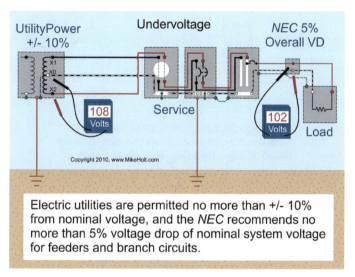

Figure 6–6

When the neutral conductor of a multiwire circuit is open, the voltage across each load on each circuit will become unstable. **Figure 6–7**

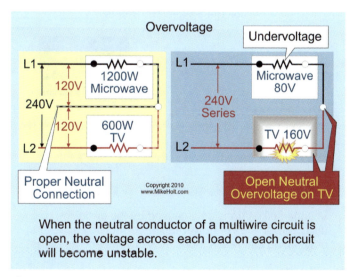

Figure 6–7

Notes

Loose conductor terminals will result in an increase in voltage drop across the terminals. This can result in a lower operating voltage at the load. **Figure 6–8**

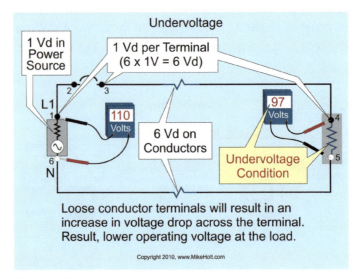

Figure 6–8

6.7 Transients (Voltage Spikes)

Transients (sometimes called voltage spikes) caused by the switching of utility power lines or power factor correction capacitors, or lightning can reach thousands of volts and amperes. **Figure 6–9**

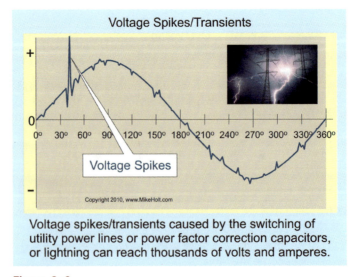

Figure 6–9

Voltage spikes (transients) produced by premises equipment such as photocopiers, laser printers, and other high reactive loads cycling off, can be in the hundreds of volts.
Figure 6–10

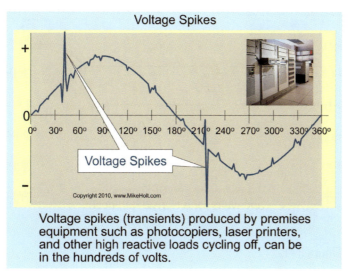

Figure 6–10

6.8 Surge Protection Devices

Surge protection devices (SPDs) protect equipment by preventing damaging transient voltage surge levels from reaching the devices they protect. Figure 6–11

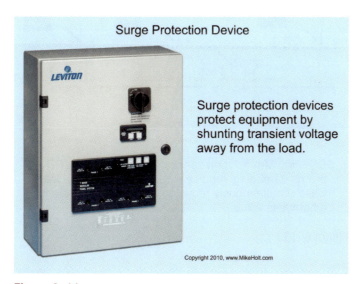

Figure 6–11

Notes

Notes

Awaiting Mode. An SPD monitors and waits for a transient to occur. **Figure 6–12**

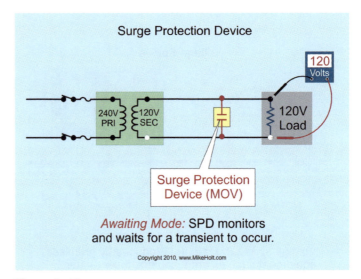

Figure 6–12

Diverting Mode. Upon sensing a transient, the SPD diverts damaging impulse current away from the load, while simultaneously reducing its resulting voltage to a harmless value. **Figure 6–13**

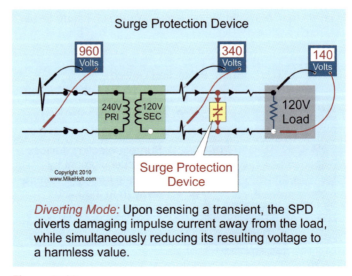

Figure 6–13

6.9 Voltage Swells

A voltage swell is caused by a short duration increase in voltage caused by a short duration decrease in current flow, such as the abrupt turning off of large loads. **Figure 6–14**

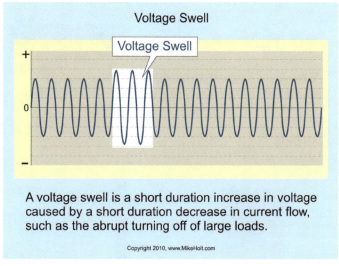

Figure 6–14

6.10 Overvoltage

Elevated voltage of more than three seconds is classified as an overvoltage condition.

The most common cause of overvoltage is an open neutral conductor. **Figure 6–15**

6.11 Unbalanced Line Voltage

Unbalanced line voltage is often caused by unbalanced phase loading of transformers or malfunctioning power factor correction devices. **Figure 6–16**

Notes

Notes

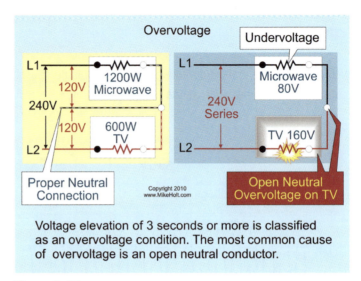

Figure 6–15

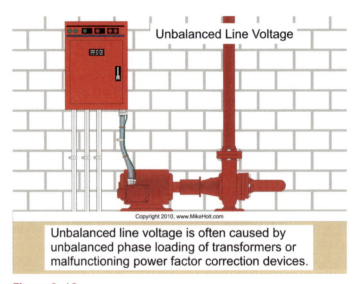

Figure 6–16

Conclusion

In this Chapter, we discussed the voltage distortion characteristics that can be introduced into the distribution system, such as flat-topping, transients, voltage notching and voltage spikes, or erratic alternating-current voltage zero crossings (nonlinear loads). Sensitive electronic equipment can fail to operate properly because of the effects of distorted nonsinusoidal-voltage waveforms.

Chapter 6 Review Questions

1. _____ voltage is produced at the source, but it is often degraded because of load variations that distort the voltage waveform.

    (a) Nonlinear
    (b) Reactive
    (c) Sinusoidal
    (d) Stray

2. Voltage _____ can occur when nonlinear loads switch on and off near the peak of the voltage waveform.

    (a) flat-topping
    (b) notching
    (c) sag
    (d) transients

3. Silicon Controlled Rectifiers (SCR) used for three-phase motor variable speed drives, large UPS systems or induction heating equipment can cause _____ in the voltage waveform.

    (a) flat-topping
    (b) notching
    (c) sag
    (d) transients

4. Voltage _____ is a short duration decrease in voltage caused by a short duration increase in current flow, such as a motor starting.

    (a) flat-topping
    (b) notching
    (c) sag
    (d) transients

Notes

5. Voltage sags of more than three seconds are classified as a(n) _____ condition.

   (a) overvoltage
   (b) undervoltage
   (c) emergency
   (d) transient

6. _____ (sometimes called voltage spikes) caused by the switching of utility power lines or power factor correction capacitors, or lightning can reach thousands of volts and amperes.

   (a) Flat-topping
   (b) Notching
   (c) Sags
   (d) Transients

7. _____ protect equipment by preventing damaging transient voltage surge levels from reaching the devices they protect.

   (a) Surge protection devices (SPD)
   (b) Power factor correction devices
   (c) Lightning protection devices
   (d) Arc-Fault Circuit Interrupters

CHAPTER 7

Voltage Window

Operating Voltage. The operating voltage of equipment must be kept within an operating window that limits overvoltages and undervoltages.

7.1 Premises Voltage Window

Electric utilities are permitted no more than a +/- 10 percent variation from nominal voltage. The *NEC* does not limit conductor voltage drop for most feeders and branch circuits, but it does recommend no more than a 5 percent voltage drop of nominal system voltage for feeders and branch circuits. **Figure 7–1**

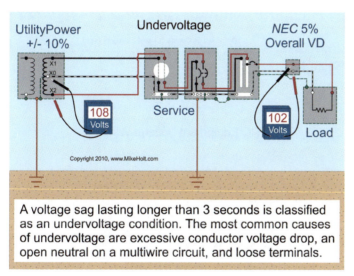

Figure 7–1

7.2 Equipment Voltage Window

The Information Technology Industry Council encourages its members to design systems to have an operating voltage window as shown in **Figures 7–2 and 7–3**. These figures show the relationship between the percentages of overvoltage or undervoltage that equipment can stand for a certain time period without sustaining damage. This is a recommended design; some equipment may be able to sustain voltages in excess of the chart values.

Notes

Notes

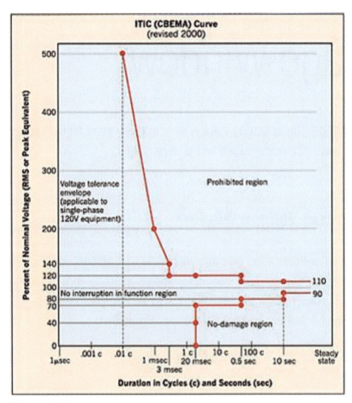

Figure 7–2

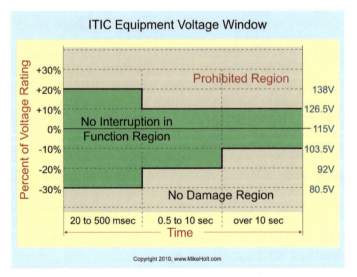

Figure 7–3

Using the proper measuring equipment, one can record voltage disturbance events for the purpose of troubleshooting equipment failures resulting from over or undervoltage. Figure 7–4 shows a sample of retained voltage versus event duration for a specific location plotted against a voltage tolerance curve called the Information Technology Industry Council (ITIC) Curve. The events are shown as green dots. Each dot represents a specific time and voltage depth. For example, one of the points is 60 cycles (1 second) and 80 percent. **Figures 7–4 and 7–5**

Notes

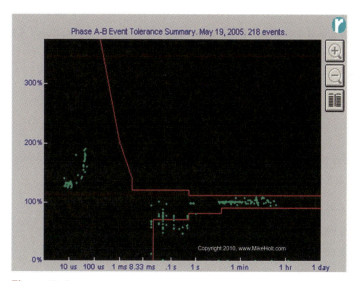

Figure 7–4

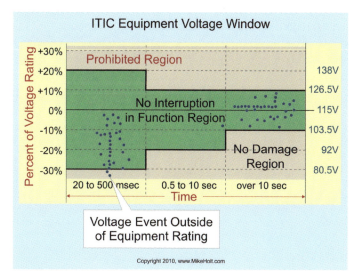

Figure 7–5

Chapter 7 — Voltage Window

Notes

Conclusion

Test equipment is available to help diagnose power quality issues by recording data to help track operating voltages.

Chapter 7 Review Questions

1. The *National Electrical Code* does not limit conductor voltage drop for most feeders and branch circuits.

    (a) True
    (b) False

2. The *NEC* suggests that the combined voltage drop for both the feeder and branch circuit should not exceed _____ of the applied voltage.

    (a) 4 percent
    (b) 5 percent
    (c) 6 percent
    (d) 7 percent

3. The Information Technology Industry Council provides recommendations to their membership regarding the proper operating voltage window.

    (a) True
    (b) False

CHAPTER 8

Electrical Noise

A rapid succession of transients on the voltage waveform induced from electrical motors and motor control devices, electric arc furnaces, electric welders, and control relays is classified as electrical noise. While electrical noise is generated from elevated harmonic frequency currents, the failure of conductor terminals can cause arcing, which can also cause electrical noise. Electrical noise is also known as "electromagnetic interference" (EMI). **Figure 8–1**

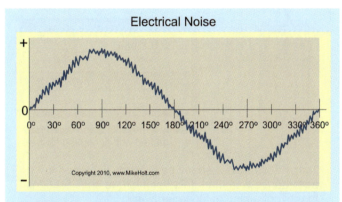

A rapid succession of transients on the voltage waveform induced from electrical motors and motor control devices, electric arc furnaces, electric welders, and control relays is classified as electrical noise.

Figure 8–1

8.1 Noise from Arcing at Terminals

Arcing at loose terminals causes electrical noise that can interfere with communications, control, data, and signal cables. When the expansion coefficient of conductors is different than the terminals, changes in temperature can result in changes at the termination. Thermal expansion and contraction between the conductors and terminals can cause the conductor to deform (or "squish"), resulting in higher terminal contact resistance and the possibility of arcing at the terminals. The increase in resistance causes a greater temperature rise (also higher voltage drop), resulting in more conductor distortion, which in turn results in greater terminal resistance.

Notes

Notes

If this cycle continues the result is greater contact resistance between the conductor and terminal permitting oxygen to bond with the conductors, thereby forming an insulating surface (copper-oxide or aluminum-oxide) and further increasing the contact resistance. All conductors must terminate in devices that have been properly tightened in accordance with manufacturer's torque specifications included with equipment instructions. Failure to torque terminals can result in excessive heating of terminals or splicing devices (due to loose connections), which can result in a fire. In addition, this is a violation of *NEC* 110.3(b), which requires all equipment to be installed in accordance with listing or labeling instructions. **Figure 8–2**

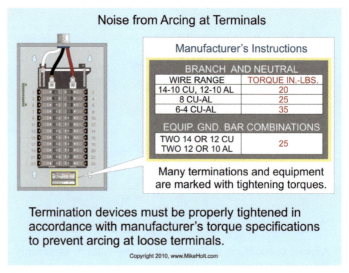

Figure 8–2

8.2 Noise from Equipment

Noise can be picked up in computer networks, communications equipment, A/V equipment, and telephone systems when harmonics are at audio or radio frequencies. **Figure 8–3**

With the increasing use of elevated frequency currents, harmonics can cause significantly greater noise on low-voltage cables, which interfere with telecommunication, video, sound, and computer systems. Harmonic currents on utility power lines can introduce noise (buzzing) on telephone lines that are on the same pole. Telecommunication circuits can be effected by frequencies as low as 180 Hz (the 3rd harmonic), and are particularly sensitive to 540 Hz frequencies (the 9th harmonic).

Figure 8–3

To reduce electromagnetic interference (noise), avoid installing communications, control, and signaling conductors near power equipment. Where possible, provide separation between power circuits and low-voltage circuits. Keep all raceways and cables away from electric-discharge lighting fixture ballasts. **Figure 8–4**

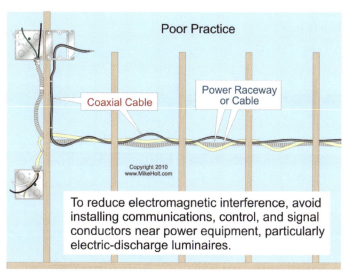

Figure 8–4

Author's Comment: Shielding of cables does not significantly reduce the noise caused by harmonic currents.

Notes

Notes

For safety reasons, the *NEC* prohibits the mixing of data and communications conductors with power and lighting circuits in the same wiring method. Remote-signaling circuits, as well as power-limited fire alarm, optical fiber cables, communications circuits, coaxial cables, and network-powered broadband communications circuits are all prohibited from being installed in the same raceway, cable, cable tray, outlet box, or similar enclosure with power or lighting conductors [725.55, 760.55, 770.133, 800.133, 820.133, and 830.133]. **Figure 8–5**

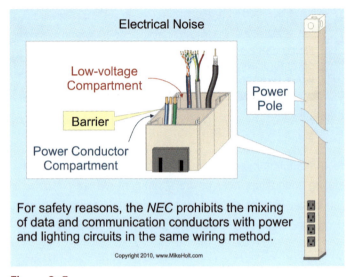

Figure 8–5

8.3 Metal Conduit Shielding

Steel conduit reduces electromagnetic fields by up to 95 percent, effectively shielding electronic equipment from electromagnetic interference. **Figure 8–6**

8.4 Shielding

Shielding of communications cables reduces radio frequency interference over 3,000 Hz; but shielding for frequency below 3,000 Hz is not effective due to the wavelength of the current. **Figure 8–7**

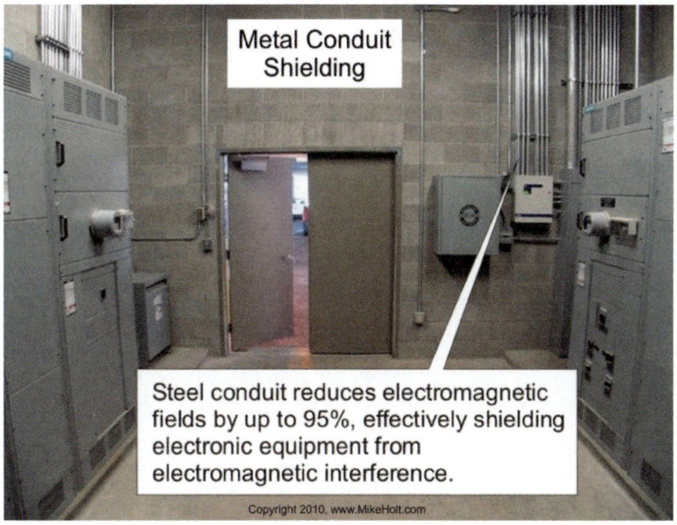

Electrical Noise — Chapter 8

Notes

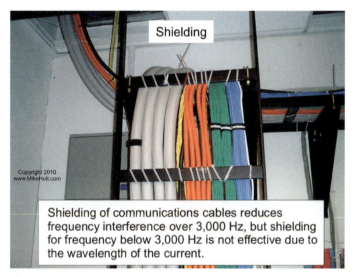

Figure 8–6

Figure 8–7

Conclusion

Noise, or electrical interference, can be caused by numerous types of equipment and by wiring installation errors or loose connections or malfunctions. Proper shielding and electrical installation practices can help reduce noise.

Notes

Chapter 8 Review Questions

1. All conductors must terminate in devices that have been properly tightened in accordance with _____ torque specifications included with equipment instructions.

    (a) the engineer's
    (b) the AHJ's
    (c) manufacturer's
    (d) the employer's

2. Failure to torque terminals is a violation of NEC 110.3(b), which requires all equipment to be installed in accordance with listed or labeling instructions.

    (a) True
    (b) False

3. To reduce electromagnetic interference (noise), avoid installing _____ conductors near power equipment.

    (a) communications
    (b) control
    (c) signal
    (d) all of these

4. Shielding of cables does not significantly reduce the noise caused by harmonic currents.

    (a) True
    (b) False

5. _____ are prohibited from being installed in the same raceway, cable, cable tray, outlet box, or similar enclosure with power or lighting conductors.

    (a) Remote-signaling circuits
    (b) Power-limited fire alarms
    (c) Optical fiber cables
    (d) all of these

6. Steel conduit reduces electromagnetic fields by up to _____ percent, effectively shielding electronic equipment from electromagnetic interference.

(a) 25
(b) 50
(c) 75
(d) 95

Notes

Notes

Grounding and Bonding

CHAPTER 9

All metal parts of the electrical installation must be bonded together and grounded to the power system earth ground. The rules for grounding and bonding are essential to provide protection from electrical shocks and fires, but also can help provide protection from transients induced by lightning or from the power distribution system.

9.1 System Grounding

All systems are either ungrounded or grounded. The advantages of grounded systems are that they help reduce equipment damage and fires from lightning and restriking ground faults. **Figure 9–1**

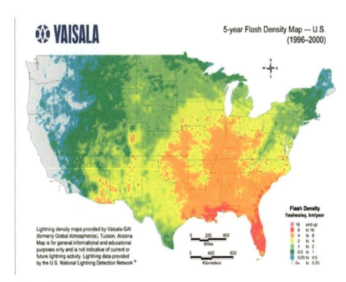

Figure 9–1

Notes

9.2 Ungrounded System

An "ungrounded system" is a system where there is no intentional electrical connection between any part of the distribution system to the earth. **Figure 9–2**

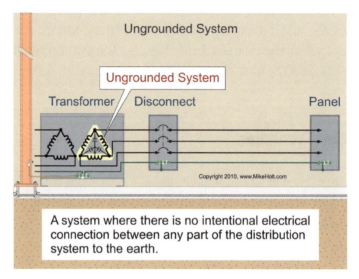

Figure 9–2

When a system is not grounded, equipment can easily be damaged from transient overvoltage caused by indirect lightning or restriking ground faults. **Figure 9–3**

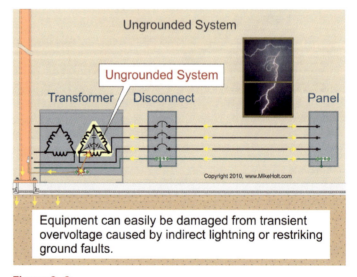

Figure 9–3

9.3 Solidly Grounded System

A "solidly grounded system" is a system that has an intentional electrical connection of one system terminal to the earth without inserting any resistor or impedance device. **Figure 9–4**

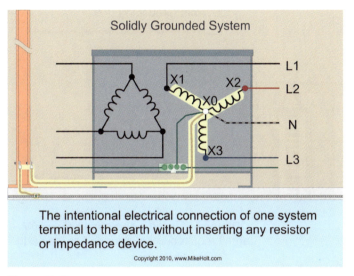

Figure 9–4

System grounding stabilizes the system voltage and reduces damage from transient overvoltage from lightning. **Figure 9–5**

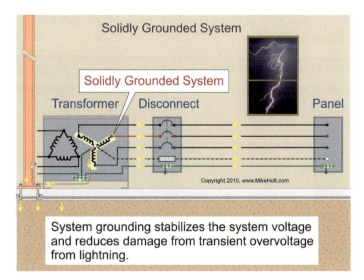

Figure 9–5

Notes

9.4 System Grounding

Premises systems operating below 50V are not required to be grounded to the earth. **Figure 9–6**

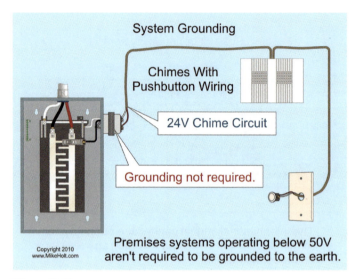

Figure 9–6

9.5 Resistance Grounded System

An impedance grounded or resistance grounded system is when an intentional electrical connection of one system terminal to the earth is made through a resistor or impedance device. **Figure 9–7**

A resistance grounded system stabilizes the system voltage and reduces damage from transient overvoltage from lightning or restriking ground faults. **Figure 9–8**

Grounding and Bonding   Chapter 9

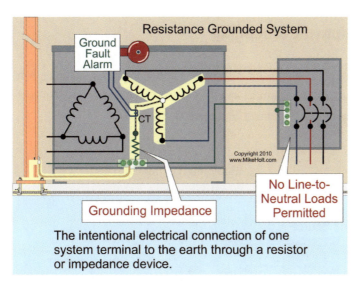

Figure 9–7

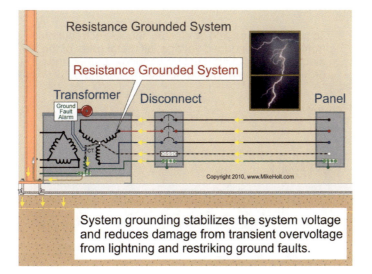

Figure 9–8

Notes

9.6 Equipment Grounding

To minimize fires and/or electric shock or electrocution from lightning induced voltage, all metal parts of the electrical installation must be bonded together and grounded to the power system earth ground. **Figures 9-9 and 9–10**

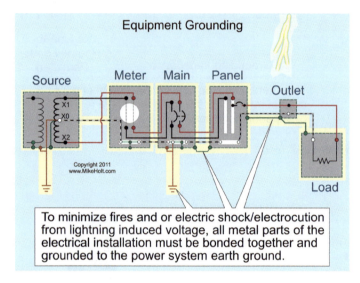

Figure 9–9

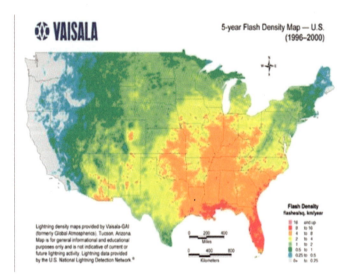

Figure 9–10

Induction from a nearby lightning strike induces dangerous transient overvoltage on metal parts which can result in a fire or electric shock and/or electrocution. **Figures 9–11 and 9–12**

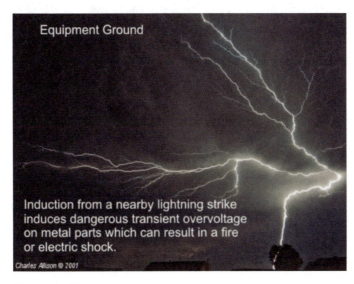

Figure 9–11

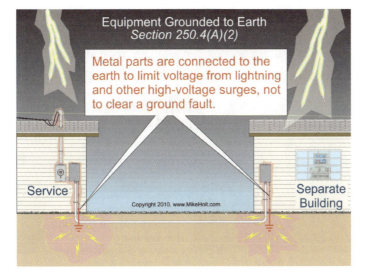

Figure 9–12

Notes

Chapter 9 — Grounding and Bonding

Notes

Failure to ground the metal parts of the electrical installation to the earth forces high-voltage transients from an indirect lightning strike to seek a path to the earth—possibly resulting in a fire and/or electric shock. **Figures 9–13 and 9–14**

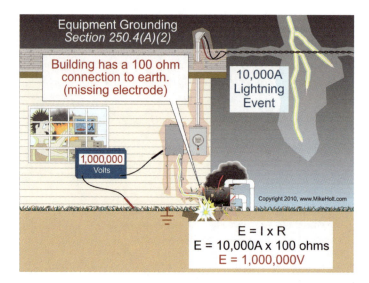

Figure 9–13

Figure 9–14

Grounding and Bonding — Chapter 9

9.7 Communications Grounding

To minimize equipment damage, fires, and electric shock and/or electrocution from lightning induced voltage, all systems must be bonded together and grounded to the same single-point ground. **Figures 9–15, 9–16, and 9–17**

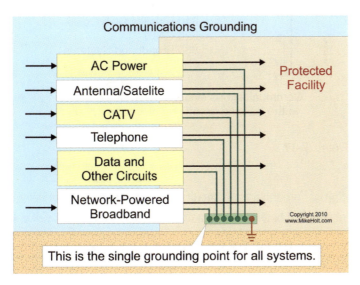

Figure 9–15

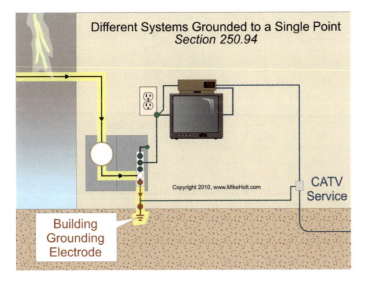

Figure 9–16

Notes

Chapter 9 — Grounding and Bonding

Notes

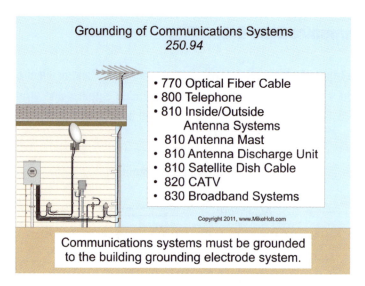

Figure 9–17

The failure of proper grounding and bonding can result in equipment failure because of a difference of potential between systems from an indirect lightning strike. **Figure 9–18**

Note: Nothing protects against a direct lightning strike.

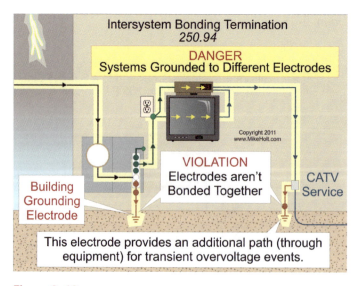

Figure 9–18

9.8 Lightning Protection System

Lightning is produced in thunderstorms when liquid and ice particles above the freezing level collide, and build up large electrical fields in the clouds. **Figure 9–19**

Figure 9–19

Once the clouds' electric fields become large enough, a giant "spark" occurs, discharging high-voltage cells between clouds to and from the earth, or from space to the earth. **Figure 9–20**

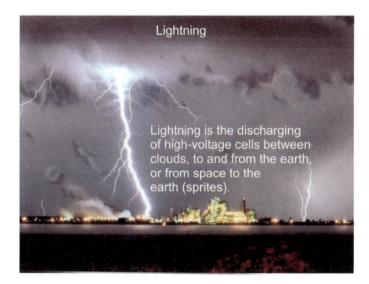

Figure 9–20

Notes

A rare positive lightning bolt originates from the top of the storm cloud rather than from the negatively charged cloud base. These massive discharges have about 10 times more current and are about 10 times longer than regular (negative) lightning. **Figure 9–21**

Figure 9–21

Lightning protection in accordance with the NFPA 780 Standard for the Installation of Lightning Protection Systems should provide some protection from damage to premises from a direct lightning strike. **Figure 9–22**

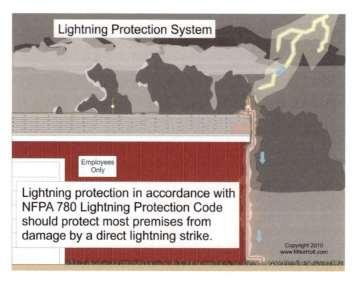

Figure 9–22

To help reduce fires in the building from side flashes, the *NEC* requires the lightning protection system to be bonded to the building grounding electrode system. **Figure 9–23**

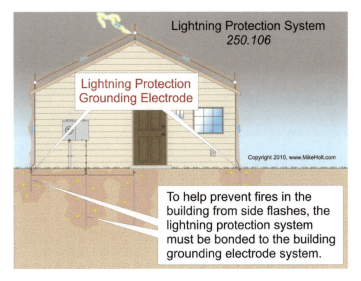

Figure 9–23

According to the *Code*, metal parts of electric equipment, when within 6 feet through air or 3 feet through dense materials, should be bonded to the lightning protection system. **Figure 9–24**

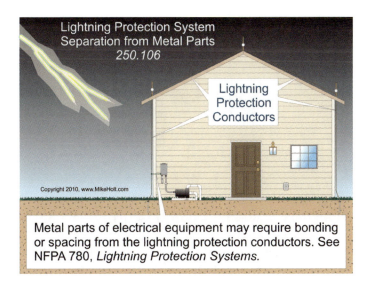

Figure 9–24

Notes

9.9 Objectionable Current

When a system neutral, and metal electrical parts are grounded to the earth at only one location, the voltage of all metal parts to the earth will be zero volts. **Figure 9–25**

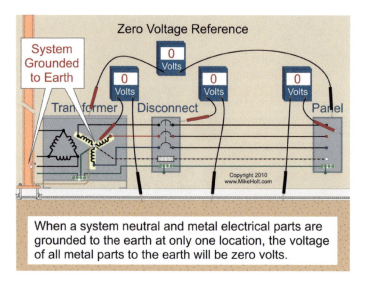

Figure 9–25

Anytime the neutral is grounded at more than one location, neutral current will flow on metal parts because of the parallel return paths to the source. The result is that the voltage from metal parts to the earth at all locations is no longer zero volts. **Figure 9–26**

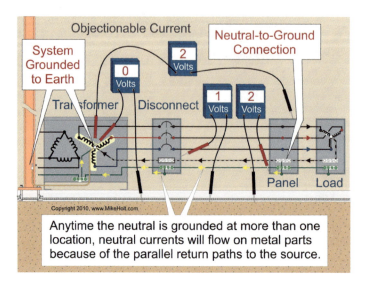

Figure 9–26

Current flowing on metal parts because of multiple neutral-to-ground connections is called "objectionable current." **Figure 9–27** Objectionable current can contribute to electrical "noise" among other problems. **Figure 9–28**

Notes

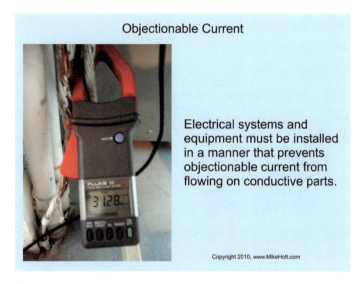

Figure 9–27

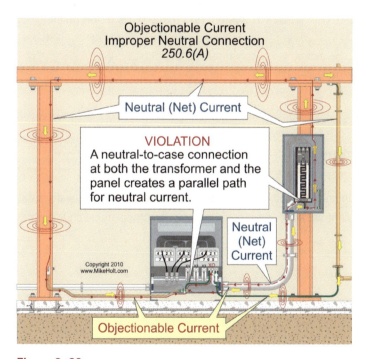

Figure 9–28

9.10 Electronic Equipment

Objectionable current flowing on metal parts of electrical equipment and building parts can cause disruptive as well as annoying electromagnetic fields which can affect the performance of electronic equipment. **Figure 9–29**

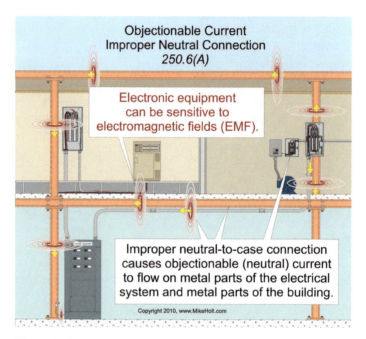

Figure 9–29

Conclusion

Proper grounding and bonding practices are essential to provide safety, and help reduce transient voltages and other sources of electrical noise on the wiring system.

Chapter 9 Review Questions

1. The advantages of _____ systems are that they help reduce equipment damage and fires from lightning and restriking ground faults.

    (a) grounded
    (b) ungrounded
    (c) redundant
    (d) three-phase

2. System grounding stabilizes the system _____ and reduces damage from transient overvoltage from lightning.

    (a) current
    (b) voltage
    (c) electrode
    (d) termination

3. An impedance grounded or resistance grounded system is when an intentional electrical connection of one system terminal to the earth is made through a _____ device.

    (a) capacitor
    (b) resistor
    (c) impedance
    (d) b or c

4. Failure to ground the metal parts of the electrical installation to the earth forces high-voltage transients from an indirect lightning strike to seek a path to the earth, possibly resulting in a _____.

    (a) fire
    (b) electric shock
    (c) resonance effect
    (d) a and/or b

Chapter 9 — Grounding and Bonding

Notes

5. To minimize _____ from lightning induced voltage, all systems must be bonded together and grounded to the same single-point ground.

    (a) equipment damage
    (b) fires
    (c) electric shock/electrocution
    (d) a, b, and c

6. According to the Lightning Protection System, metal parts of electric equipment, when within _____ feet through air or 3 feet through dense materials, should be bonded to the lightning protection system.

    (a) 2
    (b) 4
    (c) 6
    (d) 10

7. Current flowing on metal parts because of multiple neutral-to-ground connections is called _____ current.

    (a) objectionable
    (b) reactive
    (c) ghost
    (d) lost

8. Objectionable current can contribute to electrical "noise" among other problems.

    (a) True
    (b) False

Power Quality Issues

CHAPTER 10

Previous chapters discussed the basic concepts of electrical theory, such as what causes harmonic currents, and the effects of harmonic currents. This Chapter explains, in simple terms, how harmonics effect specific types of electrical equipment and what can be done to reduce those effects.

10.1 Adjustable Speed Drives

Adjustable speed drives convert incoming 60 Hz alternating-current power to direct current, which is then converted back to alternating current at the frequency defined by the speed reference of the adjustable speed drive. The alternating current to direct-current conversion is accomplished by two methods, both of which cause power quality problems:

Thyristors

Six thyristors use alternating-current power to turn six solid-state switches on and off, resulting in a current waveform that is similar to a square wave rather than a pure sine wave. This results in line voltage notching. Line voltage notching is rich in elevated frequencies that are readily propagated throughout the power distribution system. Sensitive electronic equipment, such as communications equipment, computers, or laboratory measuring and monitoring equipment is especially sensitive to signal distortion caused by line notching.

Diode Bridges

The second method of VFD alternating current to direct-current power conversion is a full-wave diode bridge with capacitors. This type of converter does not cause line voltage notching, but it does cause harmonics because it is a rectification load.

Additional zero voltage crossing caused by electronic equipment can cause disruption of equipment that uses zero voltage crossings for timing. **Figure 10–1**

Notes

Notes

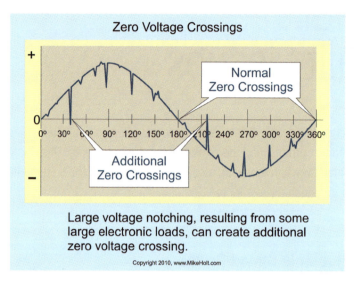

Figure 10–1

10.2 Busway Failure

A three-phase, 4-wire, busway supplying single-phase electronic equipment failed because of excessive neutral current due to odd triplen harmonic currents.

Busway neutral conductors are sized at 100 percent of the phase ampacity rating, and are vulnerable to failure when used to supply 120V or 277V nonlinear loads that produce additive triplen harmonic neutral currents. What should one do if additive triplen neutral currents exceed the neutral bus capacity? The only solution is to remove some of the nonlinear load from the bus, or to derate the bus capacity.

10.3 Capacitors

Power factor correction capacitors installed by utilities, or end users, to improve the facilities' power factor can become tuned at particular harmonic frequencies. When a capacitor becomes tuned to a specific frequency it can result in series resonance (low impedance). This can cause very high currents to flow on the conductors, which can result in overcurrent protection devices opening, and the capacitor may overheat and/or fail.

If nonlinear loads are present in a distribution system with power factor correction capacitors, the amount of true RMS current flowing through the capacitors should be measured when all the loads are on. Compare the capacitor current with the nameplate rating of the capacitor to determine if it is operating within its designed current limits.

Lighting ballast capacitors, and other capacitors in a building, can also become tuned to harmonic currents and are susceptible to current overload. Frequent failures of capacitors, tripping of circuit breakers, blowing of fuses, or overheating of conductors to capacitor loads (such as electric-discharge lighting) can be an indication of harmful harmonics in the distribution system. Harmonic currents from different loads and feeders will "migrate" to the tuned capacitor due to harmonic voltages. **Figure 10–2**

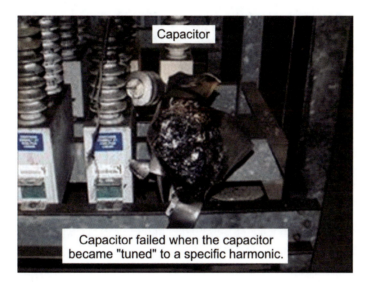

Figure 10–2

Notes

Notes

10.4 Conductor Failure

Conductor insulation can fail due to excessive neutral current because of improper wiring. **Figure 10–3**

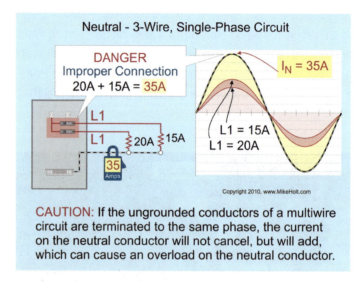

Figure 10–3

Existing feeder neutral conductors of a 4-wire, three-phase system can fail because of additive odd triplen harmonic currents.

The *National Electrical Code* conductor ampacities listed in Table 310.16 are based on not more than three current-carrying conductors in a raceway or bundle of cables, installed in an ambient temperature of 86°F, at the fundamental frequency of 60 Hz. Conductors heat up beyond normal I^2R levels because of eddy currents and skin effect that are the result of harmonic currents operating at frequencies higher than 60 Hz.

Since skin effect produces additional heating and the apparent reduction in conductor cross-sectional area (which increases resistance), the current-carrying capacity (ampacity) of the conductor must be reduced. Section 310.10 of the *NEC* requires that conductors must not be used in any way that the temperature rating of the conductor will be exceeded. Because it is not possible to precisely establish the heating effects of eddy currents and skin effect, there is presently no method of adjusting conductors for the effects of harmonic currents. Some industry experts suggest that the load should be limited to 70 percent of the conductor rating, or the phase conductors be sized one size larger than required by the *Code*.

10.5 Circuit Breakers

Circuit breakers generally have two types of sensing devices, the short-circuit/ground-fault (magnetic) device, and the long-time overload (thermal) device. Peak-sensing electronic circuit breakers have been known to trip prematurely when supplying high crest factor loads. **Figure 10–4**

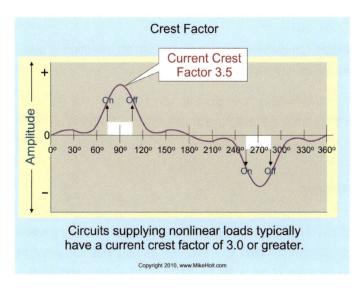

Figure 10–4

Although harmonic currents and inductive heating can cause circuit breakers to prematurely trip, the most likely reason a circuit breaker trips is because of overload.

Magnetic Trip Circuit Breakers. Circuit breakers that supply nonlinear loads do not respond magnetically as calibrated due to the fact that the inductive reactance of an inductor (coil) changes with frequency [$X_L = 2\pi f L$]. This can result in a noncoordinated overcurrent protection scheme, which can cause an undesirable power outage.

Thermal Trip Circuit Breakers. Increased current frequencies due to harmonics result in additional skin effect and possible inductive heating of circuit breakers. Additional heating from phase conductors and metal enclosures also contribute to the internal heating of circuit breakers, and can result in premature operation at less than trip or fault-level currents. Some experts in the industry suggest that loads should not exceed 70 percent of the overcurrent device rating to allow for extra heating caused by high harmonic currents. The *NEC* requires that circuits not be loaded more than 80 percent for continuous loads (3 hours or more). [210.20(a) and 215.3]

Notes

10.6 Electric-Discharge Luminaires

Energy savings of 25 to 50 percent can be achieved by using electronic ballasts instead of standard electromagnetic ballasts. Electronic ballasts start and regulate fluorescent lamp voltages using electronic components rather than the traditional core-and-coil assembly. While electromagnetic ballasts operate at the fundamental 60 Hz, electronic ballasts convert 60 Hz input voltage to an output voltage operating at a frequency that can be between 20,000 and 60,000 Hz.

Electronic ballasts can cause excessive neutral current, resulting in neutral conductor failure because of additive triplen harmonic currents.

Solution—Ensure luminaires have passive inductive-capacitive filters to reduce harmonics. Newer electronic ballasts have greatly reduced harmonic distortion.

10.7 Dimmers

Some manufacturers of electronic dimmers require that a separate neutral conductor be used to supply each device, or that the neutral conductor of a 4-wire multiwire branch circuit be sized at twice the ampacity of the phase conductors. This requirement is because of the additive triplen harmonic currents. Electronic dimmers are basically direct-current loads that generate harmonics. Electronic dimmers can fail because of overvoltage or undervoltage resulting from an open neutral. Solution—Install a separate neutral conductor for each circuit. **Figure 10–5**

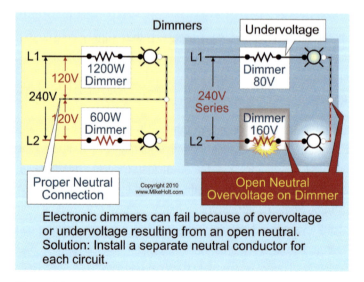

Figure 10–5

10.8 Overcurrent Protection Devices

Overcurrent protection devices respond to True-RMS current, and in most cases, will protect the phase conductors against overloads that result from nonlinear loads. However, overcurrent protection devices cannot protect against eddy currents, skin effect, induction heating, and circulating triplen currents in the primary of delta/wye transformers. Overcurrent devices can open at currents below the device calibration rating because of unanticipated heating from eddy currents, skin effect, and hysteresis losses. **Figure 10–6**

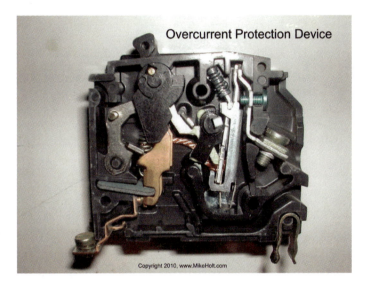

Figure 10–6

Fuses are calibrated based on 60 Hz current. Currents at higher frequencies can "blow" the fuse at currents below the fuse calibration rating because of unanticipated inductive heating from eddy currents, skin effect, and hysteresis losses. In effect, a 100A fuse supplying a nonlinear load could blow at some value less than 100A RMS.

10.9 Generators

Generators and other electromagnetic devices (transformers, coils, and so forth) are designed to operate at a fundamental voltage frequency of 60 Hz. Electromagnetic devices are vulnerable to overheating from increased eddy currents in the core and conductors, and skin effect losses, hysteresis losses, and circulating harmonic currents

Notes

Notes

in the primary. Backup generators have overheated and failed because of excessive eddy currents, hysteresis losses, and circulating triplen harmonic currents. **Figure 10–7**

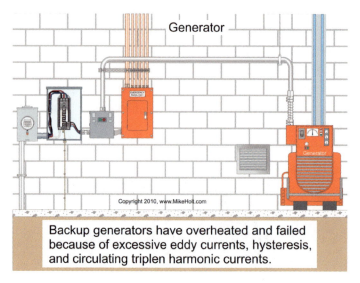

Figure 10–7

In addition to overheating, harmonic currents from nonlinear loads have been known to distort the voltage waveform, which can cause interference with the generator's voltage regulator. If the peak current on the generator exceeds the generator's magnetic capacity, the output voltage waveform can flatten excessively resulting in insufficient RMS voltage for proper equipment operation. **Figure 10–8**

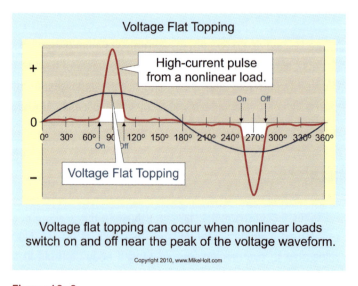

Figure 10–8

Generators can be constructed to accommodate the heating effects of harmonic currents by the use of higher temperature-rated conductors, higher quality steel, reducing the thickness of laminations, and by changing the winding configuration of the generator rotor.

However, generator manufacturers are not specifically constructing, marking, or listing their generators for nonlinear loads. What most generator manufacturers recommend in relation to power quality issues is to simply increase the generator size by a factor of 1.40 or limit the load to 70 percent of the generator's rating.

10.10 Laser Printers

A laser printer on the same circuit with data processing equipment has caused data errors because of a voltage sag when the "print-fusing heater" cycles on and off.

Solution—Don't have electronic loads on the same circuit with high-current nonlinear loads.

10.11 Light Flicker

An electrical contractor was sued by a homeowner because their incandescent lights flickered when the air-conditioning started. Customers may become alarmed if the lights flicker. Sometimes there are voltage drop issues, distance from the utility transformer factors, and other issues which are not safety related and which are not easily remedied.

Solution—Not much… **Figure 10–9**

Notes

Notes

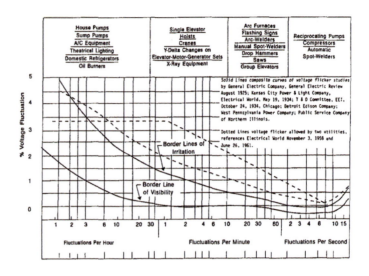

Figure 10–9

10.12 Modular Office Furniture

Prewired office furnishings supplied with a multiwire circuit caught on fire because of excessive neutral current caused by triplen harmonic currents.

Solution—Order furniture with oversized neutral or separate neutral conductor for each circuit.

10.13 Motors

To conserve electricity, and for more precise control, adjustable-speed drive motors are often used in saw mills, irrigation systems, lathes, woodworking equipment, extrusion processes, and other industrial processes. Adjustable-speed drive motors operate on three-phase rectified, 6- or 12-pulse power, which permits a wide range of speed control and energy consumption. The switching power supplies for these types of loads produce significant 5th harmonic (300 Hz) currents, as well as notching disturbances.

Alternating-current motors can overheat because of (1) reduced voltage (voltage drop), (2) unbalanced voltage, (3) inductive heating (eddy currents, hysteresis losses in the core, and skin effect), and (4) negative magnetic field rotation. If an induction motor is overheating, or unexplained burnouts are being experienced, the first thing to do is verify that the motor is operating within its nameplate current rating. Also, be sure the motor is operating at, or above, the correct RPM rate and check the voltage to ensure it is not unbalanced. If the motor is drawing current according to the nameplate and still overheating, the likely culprit is harmonic currents.

Unfortunately, one cannot determine if harmonics are causing the motor failures without analyzing the power system for harmonics with a spectrum harmonic analyzer (at a cost of several thousand dollars).

In addition to inductive heating, alternating-current motors can fail because of overcurrent resulting from negative magnetic field rotation. Alternating-current motors depend on a 60 Hz positive voltage rotating magnetic field to turn the motor in a predetermined rotation. Harmonic voltages of the 5th, 11th, 17th, 23rd (and so on) levels have a negative field rotation that attempts to run motors backwards, countering the positive applied voltage. Negative field rotation caused by the 5th and 11th harmonics, in particular, reduce motor efficiency and dramatically increase motor operating temperature. Motor life can be reduced because negative sequence 5th and 11th harmonics attempt to drive three-phase alternating-current motors in reverse. **Figure 10–10**

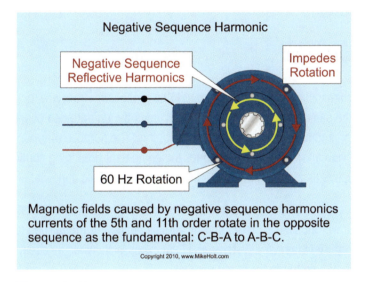

Figure 10–10

Solution—Install a passive LC filter tuned to the offending harmonic frequency.

10.14 Panelboards

The most significant problems associated with harmonic currents within panelboards are neutral conductor terminal connection failures because of excessive neutral current resulting from unbalanced phase currents and additive triplen harmonic currents. A sign of excessive harmonic neutral current is discoloration of the neutral conductor near its termination.

Notes

There are UL-listed panelboards available for nonlinear loads, known as "NL Panelboards." The neutral terminal is designed to carry current of up to 200 percent of the phase bus current rating. These panelboards are made in ampere ranges from 100A to 400A for 277/480V, three-phase, 4-wire and 240 VAC. The only difference between a standard panelboard and a type NL Panelboard is that the neutral terminals are larger; they cost more (at least 30 percent more in a recent check on common sizes), but they are required if neutral terminal failure is to be avoided unless a harmonic filter is installed to protect them.

Metal panelboard enclosures can become very hot because of the inductive and hysteresis heating of steel enclosures due to the presence of harmonics. Panelboards are built to not be mechanically resonant at 60 Hz (this means that the panelboard will not make a buzzing sound at 60 Hz). Elevated-frequency harmonic currents can cause panelboards to vibrate and emit a buzzing sound. Although this is not a hazard, it can be annoying.

10.15 Raceways

The *NEC* requires all conductors of a circuit to be in the same raceway, or placed close together in the same trench [300.3(b)]. This includes the ungrounded conductors, neutral conductor, and equipment grounding conductors. The purpose of grouping conductors is to provide a low-impedance path for the equipment grounding conductor, and to prevent inductive heating of metal parts through magnetic field cancellation.

When conductors are grouped together, the magnetic fields neutralize each other resulting in lower induced currents and hysteresis heating of the ferrous (iron or steel) metal raceway. Steel raceway systems will heat excessively when all circuit conductors are not grouped together in the same raceway.

Solution—Install all wiring methods in accordance with the *Code*. **Figure 10–11**

Induction heating occurs when the magnetic fields of the additive triplen neutral currents expand and collapse at a frequency of 180 Hz within the ferrous metal raceway, resulting in eddy currents and hysteresis heating. When the raceways and enclosures heat up, the circuit conductors are also heated which further reduces their ampacity.

10.16 Transformers

Standard transformers are designed to operate within their rated operating temperature rise when loaded at not more than their rated kVA, and if the nonlinear load they supply is no more than 5 percent of the total load. If the nonlinear current load

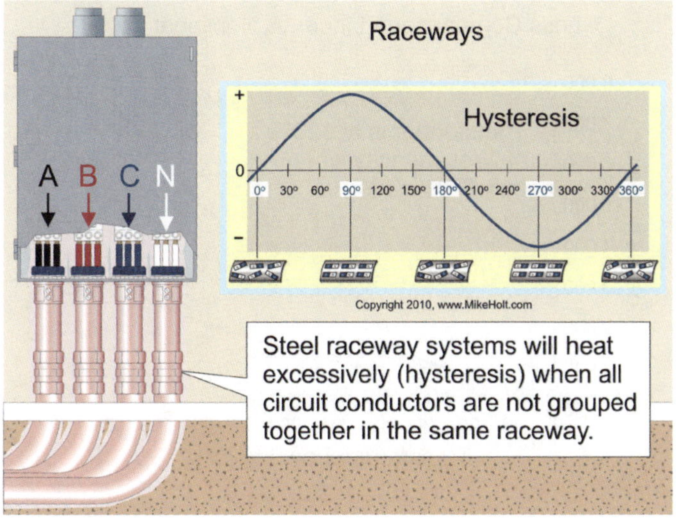

Figure 10–11

of a transformer exceeds 5 percent, the elevated-frequency reflective harmonic currents can overheat the transformer windings resulting in transformer failure, even if the transformer is loaded at less than its kVA rating. In addition to transformer failures, harmonic currents can cause significant voltage waveform distortions of the voltage supply, which can effect other electrical equipment on the power distribution system.

Transformers often fail because of inductive heating from circulating eddy currents, hysteresis losses, and additive triplen harmonic currents. When a transformer is carrying nonlinear loads, it can reach 100 percent of its heating capacity at 60 percent of its rated load. **Figures 10–12 and 10–13**

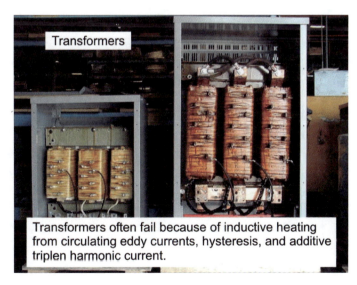

Figure 10–12

Notes

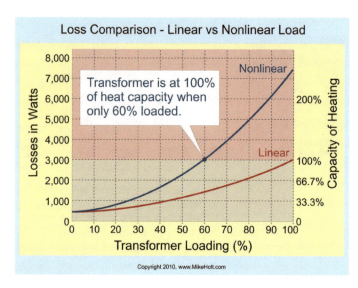

Figure 10–13

Inductive Heating. Inductive harmonic heating to the transformer winding is in proportion to the square of the harmonic current frequency; 3rd harmonic inductive heating is nine times the percentage of its fundamental current heating; and at the 5th harmonic, inductive heating is 25 times its percentage of the fundamental current heating. Because of harmonic current inductive heating, transformers that supply nonlinear loads in excess of 5 percent of the total transformer load rating must be derated or replaced with transformers specially designed to supply the nonlinear loads.

Transformers have magnetic, as well as thermal characteristics. If momentary peak currents of nonlinear loads exceed the transformer's peak current rating, the transformer core can become magnetically saturated. This results in a distorted (flat) voltage waveform output. Distorted voltage waveforms affect various types of equipment differently.

K-Rated Transformers. K-rated transformers are designed to handle the heating effects resulting from nonlinear loads. They do not remove harmonic distortion from the system.

Underwriters Laboratories has established ratings of 1, 4, 9, 13, 20, 30, 40, and 50 as standard K-factor ratings. A higher K-rating represents the capability to withstand higher harmonic content, with K4 and K13 being the most frequently specified.

K-rated transformers include neutral lugs that allow for 200 percent neutral connections, smaller parallel secondary windings, and transposed delta winding conductors.

Naturally, K-rated transformers cost more than non-K-rated transformers. It is impossible to determine the K-factor loading for new tenant space because the building's load characteristics cannot be anticipated. A manufacturer of transformers can provide guidance in sizing K-rated transformers.

10.17 Relays

It is impossible to generalize the behavior of any particular relay in response to harmonic currents without testing. This applies to static, solid-state, induction, plunger, or clapper-type relays. As the percentage of harmonics on a system increases, installations that initially caused no trouble may experience nuisance operation and relay failure due to inductive heating.

10.18 Transfer Switches

Transfer switches are used in many facilities to transfer electric power from the utility system to backup generators. UL Standard 1009 requires the neutral terminal of a switch to be sized at 100 percent of the phase current rating. Additive triplen harmonic currents can cause the neutral conductor to carry current in excess of the phase current, resulting in the neutral terminal and switching mechanism to fail because of overload. Neutral terminal failure can destroy thousands of dollars of electrical equipment because of overvoltage and undervoltage.

Solution—Increase the neutral size and ensure that the terminals of the transfer switch can accommodate a larger neutral conductor.

10.19 Uninterruptible Power Supplies (UPS)

Static and rotary uninterruptible power supplies (UPS) convert alternating current to direct current, and then back to alternating current. This switching power supply can generate tremendous 5th, 7th, 11th, and 13th order harmonic currents back into the alternating-current distribution system. Today, most of the newer UPS systems have inductive capacitive (LC) filters, which reduce the reflective harmonic currents. UPS systems without LC filters can have a total harmonic distortion (THD) of as much as 35 percent. Electronically controlled equipment in one new facility locked up and experienced repeated data errors because of severe phase voltage notches caused by a UPS system.

Notes

Notes

Conclusion

This textbook has provided a comprehensive overview of the important subject of power quality. It concentrated on the phenomenon of harmonic currents caused by nonlinear loads, and the effects these harmonics have on premises wiring systems and electric utilization equipment.

Although harmonics caused by nonlinear loads are a serious and growing problem in today's electric power systems, designers and specifiers have yet to develop a comprehensive approach to solving the problem. The *National Electrical Code* contains many provisions dealing with the problem of harmonics, most of them developed by a special "Ad Hoc Subcommittee on Nonlinear Loads" for the 1996 edition of the *Code*. But in general these references are informational Notes, rather than mandatory *NEC* rules intended to minimize the safety effects of this growing problem. Technical understanding of the causes and consequences of harmonics is not yet good enough to permit the development of a comprehensive solution in the *Code*.

The "harmonic profile" of any building is unique. Specific solutions to the problems described in this textbook must be based on analysis and study to determine the frequency and magnitude of harmonics, their sources, and their effects on the power system. However, although diagnostic tools such as spectrum analyzers, recording power meters, and special software can be used to study power system characteristics, the electrical industry has not yet defined standard methods for testing and troubleshooting power quality problems caused by harmonics.

A number of vendors are now marketing low-harmonic alternating-current/direct-current power supplies, and these are being installed by some electronic equipment manufacturers. Likewise, utilities and standards developers such as IEEE and NEMA are studying the harmonics problem. Future changes to the *National Electrical Code* will eventually spur improvements in equipment design, listing, and installation. These factors will be necessary if designers and installers are to have more options.

Because it is impossible to obtain specific knowledge of a building's harmonic profile before the electrical system is built and the utilization equipment installed, the measures that can be taken in the meantime are few and basic:

1. Assume all neutral conductors to be current-carrying and apply the ampacity adjustment factors from 310.15(b)(2)(a) of the *NEC*.

2. Size all common neutral conductors at 200 percent of the phase current.

3. Balance phases as closely as possible to minimize neutral current.

4. Install K-factor rated transformers and NL-rated panelboards or use harmonic filters.

The *National Electrical Code* addresses harmonics and nonlinear loads in the following sections:

- Article 100 Definition "Nonlinear load" and its Note
- 210.4(a) Note
- 220.61(c)(2) and its Note, Exception 4, Note item (2)
- 310.4 Exception 4, Note
- 310.15(b)(4)(c)
- 368.258
- 400.5(b)
- 450.3 Note 2
- 450.9 Note 2
- Table 520.44 Note
- Example D3(a) in Annex D

Chapter 10 Review Questions

1. Busway neutral conductors are sized at _____ of the phase ampacity rating, and are vulnerable to failure when used to supply 120V nonlinear loads that produce additive triplen harmonic neutral currents.

    (a) 25 percent
    (b) 45 percent
    (c) 50 percent
    (d) 100 percent

2. When a capacitor becomes tuned to a specific frequency, it can result in series resonance (low impedance). This can cause very high currents to flow on the conductors, which can result in _____.

    (a) overcurrent devices opening
    (b) the capacitor overheating
    (c) the capacitor failing
    (d) all of these

Chapter 10 — Power Quality Issues

Notes

3. If linear loads are present in a distribution system with power factor correction capacitors, the amount of true RMS current flowing through the capacitors should be measured when all the loads are on.

    (a) True
    (b) False

4. The *National Electrical Code* conductor ampacities listed in Table 310.16 are based on not more than three current-carrying conductors, bunched in an ambient temperature of_____, at the fundamental frequency of 60 Hz.

    (a) 90°F
    (b) 86°F
    (c) 72°F
    (d) 68°F

5. Since skin effect produces additional heating and the apparent reduction in conductor cross-sectional area (which increases resistance), the current-carrying capacity (ampacity) of the conductor must be reduced.

    (a) True
    (b) False

6. The *NEC* requires that circuits not be loaded more than _____ for continuous loads (3 hours or more).

    (a) 50 percent
    (b) 60 percent
    (c) 70 percent
    (d) 80 percent

7. Energy savings of _____ percent can be achieved by using electronic ballasts instead of standard electromagnetic ballasts.

    (a) 5 to 10
    (b) 15 to 20
    (c) 25 to 30
    (d) 25 to 50

8. _____ can cause excessive neutral current, resulting in neutral conductor failure because of increased harmonic currents.

   (a) Fluorescent ballasts
   (b) Electronic ballasts
   (c) Energy-saving ballasts
   (d) Electromagnetic ballasts

9. Fuses are calibrated based on 60 Hz current. Currents at higher frequencies can "blow" the fuse at currents below the fuse calibration rating because of unanticipated inductive heating from _____.

   (a) eddy currents
   (b) skin effect
   (c) hysteresis losses
   (d) all of these

10. Generators can be constructed to better withstand the heating effects of harmonic currents by _____.

    (a) installing higher temperature-rated conductors
    (b) the use of higher quality steel
    (c) reducing the thickness of lamination
    (d) all of these

11. What most generator manufacturers recommend is simply to _____ the generator size or limit the load to some fraction of the generator's rating (adjust the generator).

    (a) increase
    (b) decrease
    (c) double
    (d) triple

12. Alternating-current motors can overheat because of _____.

    (a) reduced voltage
    (b) unbalanced voltage
    (c) inductive heating
    (d) all of these

Notes

Chapter 10 — Power Quality Issues

Notes

13. If an induction motor is overheating, or unexplained burnouts are being experienced, the first thing to do is verify that the motor is operating within its nameplate current rating.

    (a) True
    (b) False

14. The most significant problems associated with harmonic currents within panelboards are neutral conductor terminal connection failures because of excessive neutral current resulting from _____.

    (a) unbalanced phase currents
    (b) additive triplen harmonic currents
    (c) nonlinear load overloads
    (d) both a and b

15. There are UL-listed panelboards available for nonlinear loads, known as NL Panelboards in which the neutral terminal is designed to carry up to _____ of the phase bus current rating.

    (a) 50 percent
    (b) 75 percent
    (c) 100 percent
    (d) 200 percent

16. Induction heating occurs when the magnetic fields of the additive triplen neutral currents expand and collapse at a frequency of _____ within a ferrous metal raceway, resulting in eddy currents and hysteresis heating.

    (a) 20 Hz
    (b) 60 Hz
    (c) 100 Hz
    (d) 180 Hz

Power Quality Issues — Chapter 10

17. Standard transformers are designed to operate within their rated operating temperature rise when loaded not more than their rated kVA, and if the nonlinear load they supply is no more than _____ of the total load.

    (a) 5 percent
    (b) 10 percent
    (c) 15 percent
    (d) 20 percent

18. The primary reason of transformer failure is: _____.

    (a) Inductive heating from circulating primary harmonic currents.
    (b) Increased resistive heating because of increased eddy currents and skin effect in conductors.
    (c) Increased inductive core heating from eddy currents and hysteresis losses.
    (d) all of these

19. At the 5th harmonic, inductive heating is _____ times its percentage of the fundamental current heating.

    (a) 9
    (b) 15
    (c) 25
    (d) 35

20. It is possible to determine the transformer K-factor loading for new tenant space.

    (a) True
    (b) False

21. It is impossible to generalize the behavior of any particular relay in response to harmonic currents without testing.

    (a) True
    (b) False

Notes

Chapter 10 — Power Quality Issues

Notes

22. UPS systems without LC filters can have a total harmonic distortion (THD) of as much as _____.

   (a) 35 percent
   (b) 45 percent
   (c) 55 percent
   (d) 65 percent

Technical References

The Institute of Electrical and Electronics Engineers (IEEE) provides publications that contain recommended practices in the area of harmonic currents.

American Power Conversion is a good reference for further research at www.APC.com

Notes

Notes

Notes

Notes

- -$4,400 LMCU cc   pay by 20th
- -$4,800 Fernando
- +$3,327 LMCU   -$9,715
- +$4,765 LMCU   +$18,092
- -$615 pay by 9th   -$1,623

- quoted for installing crown molding - $160?
- materials were already provided for drywall - $200